Praise for *An Insider's Guide to Mentoring the Fire Officer*

"This is a must read for all aspiring officers, officers climbing the leadership ladder, or those at the summit. In Chief Shapiro's book, An Insider's Guide to Mentoring the Fire Officer, pages come alive with scenes from mentoring discussions straight from the fireground, and firehouse kitchen tables. Gripping, sad, heartfelt, funny, surprising, informative: it's all very relatable. His real-life descriptions of firefighter mentoring are how all firefighters should lead. With fire service life lessons and wisdom from a savvy, experienced street boss, this book is for all to learn from and benefit."

—Michael Ciccarelli
Deputy Fire Chief (ret.), Hartford (CT) Fire Department
Instructor (ret.), CT Fire Academy

"Chief Leigh Shapiro's book is a genuine, invaluable resource, shaped by decades of practical real-world and hands-on experiences, designed to impart knowledge and valuable insight that fosters personal and professional growth as a leader. This book will help one navigate the complexities of fire service leadership and mentoring, providing humbling advice and experience-based wisdom for challenging situations and tough decisions. This book is a testament to Leigh's desire to motivate and inspire to lead with strength, confidence, purpose, and integrity."

—Jimmy Davis
Captain, Chicago (IL) Fire Department

"An Insider's Guide to Mentoring the Fire Officer is outstanding! Deputy Chief Shapiro brings a wide variety of experience with significant depth to help new and seasoned officers strengthen their ability to lead, communicate, and mentor. It is well written, with great details and relatable incidents that any line or chief officer can relate to. There are many books out there that help leaders lead, but this one helps fire officers lead in an extraordinary business where lives are collateral, and decisions have significant consequences."

—Walter Lewis
Assistant Chief, Orlando (FL) Fire Department

An Insider's Guide to Mentoring the Fire Officer

AN INSIDER'S GUIDE TO MENTORING THE FIRE OFFICER

LEIGH H. SHAPIRO

Fire Engineering Books

Disclaimer

The recommendations, advice, descriptions, and methods in this book are presented solely for educational purposes. Photos are for instructional purposes only. Always wear the proper level of approved personal protection equipment (PPE) when conducting training drills and operating at incidents. The author and publisher assume no liability whatsoever for any loss or damage that results from the use of any of the material in this book. Use of the material in this book is solely at the risk of the user.

Copyright © 2025 by
Fire Engineering Books & Videos
110 S. Hartford Ave., Suite 200
Tulsa, OK 74120 USA

800.752.9764
+1.918.831.9421
info@fireengineeringbooks.com
www.FireEngineeringBooks.com

Executive Vice President: Eric Schlett
Vice President, Group Publishing: Amanda Champion
Acquisitions: David Rhodes, Diane Rothschild
Director, eLearning and Books: Starlet Franz
Production Manager: Tony Quinn
Book Designer: Robert Kern, TIPS Publishing Services, Carrboro, NC
Cover Designer: Brandon Ash

Library of Congress Cataloging-in-Publication Data Available on Request

ISBN: 9781593706043

All rights reserved. No part of this book may be reproduced, stored in a retrieval system, or transcribed in any form or by any means, electronic or mechanical, including photocopying and recording, without the prior written permission of the publisher.

Printed in the United States of America

1 2 3 4 5   28 27 26 25

This book is dedicated to my wife Malena, who has been by my side for the whole ride. And for Kevin.

Contents

Preface xi
Acknowledgments xiii
Introduction xv

1. Alarm of Fire! 1
   Case Study 6
   Questions 6

2. Mentoring and Coaching the Fire Officer: Pearls of Wisdom . . . 7
   New Duties and Responsibilities 9
   Critical Thinking for Fire Officer Development 15
   Mentoring and Coaching 25
   Personnel Issues 41
   Sexual Harassment 48
   Case Study 53
   Questions 54

3. Succession Planning: Promotional Preparation
   and Advancement 55
   Case Study 65
   Questions 67

4. Decision-Making: Know Your Job! 69
   Indecision Is No Decision, So Make a Decision 74
   Which Side Are You On? 94
   Prepared versus Ready: Fundamental Differences 98
   Case Study 106
   Questions 107

5 The Incident Commander: Points to Enhance Skillsets 109
The Street Boss 111
The Dance Card: Fireground Accountability for the Incident
  Commander, Simplified! 119
Simplified Format for Incident Management 128
High-Rise Fires 133
Technical Rescue 135
Case Study 137
Questions 140

6 Fireground Behaviors 141
Factors Affecting Behavior 143
Overcoming Three Common Fireground Failures 154
Interior Fire Attack: Obsolete or Indispensable? 158
Case Study 163
Questions 165

7 Analysis of an After-Action Report 167
Case Study 175
Questions 175

8 The Incident: The Impact of a Fire Service Line-of-Duty Death 177
The Ultimate Sacrifice 178
The Aftermath 179
Evidence Collection 181
The Unexpected 192
The Notification 196
The Impact 198
What's Left 200
Case Study 204
Questions 205

9 Final Thoughts 207
Case Study 209
Questions 210

Index 211

Preface

Ask anyone within a given profession and they will gleefully explain that a mentor, or the receivership of information, is the mortar which binds the foundational building blocks of training, education, and experience. For without that coaching, that guidance, the tasks at hand become more difficult and at times confusing and overwhelming. To truly appreciate the nuances, the big picture, and the why of our chosen vocation as firefighters, both career and volunteer professionals, we must first grasp that helping hand and steadfast dependability of a mentor, a coach, or simply someone to turn to for advice. The learning environment is ever present, whether it's a fleeting moment at an incident or event, or throughout your entire career's experience. It is because of this that the information and guidance provided within the confines of this text seeks to speak with a motivating, informative, steady voice of a mentor—always available, trustworthy, and dependable. Throughout my 28-year career as a firefighter, I often reached out to those who I knew would communicate that critical information to me when I needed it, and even sometimes when I thought I didn't. Whether it was one or two trusted individuals, or even those whom I had no relationship with other than in a professional capacity, I was never timid about asking questions and embracing the comprehensive learning of processes and components of my chosen profession. I have often said after I retired from Hartford Fire (HFD) that with all my years of training, education, and experience, I metaphorically had a Ferrari parked in my driveway, and since I am no longer working for the HFD, I had no place to drive it. So, what does one do in this dilemma? Simple: I built a road! I am now able to give back to my beloved fire service in many ways, with many tools at my disposal. This book is one of those tools in my toolbox.

Acknowledgments

None of what I have accomplished in my life, beginning with the inspiration to be a firefighter, would have been possible without the love and support of my family. Starting with my grandmother, aunt, and parents, and continuing through to my own wife and children, they have all been right there beside me for the good and the bad and will always be my steadfast support system. From the moment I met my future wife in high school, she has been on this fire service journey the whole way. My wife has always stood by her words of strength and encouragement when she would reaffirm to me that I myself was the only limitation to my success! Each of my three children have validated and reaffirmed my career path in their own distinct way: My daughter served on a city ambulance as an emergency medical technician for over 12 years and is now an emergency room trauma nurse at the state's busiest emergency room in the capital city; my two sons are both Hartford Firefighters—one serving on the city's busiest truck company, the other a lieutenant on the city's rescue company. The friendships I have made and colleagues I have worked with throughout my career and beyond are invaluable and have greatly contributed to my success. The many mentors, officers, and acquaintances I have been fortunate to learn from have provided me with the life path that has served me well and has greatly contributed and influenced the materials found in this book. Even when I doubted myself and the worthiness of this material, there were those who saw beyond the immediate and proclaimed the value and utility of passing along my years of collected knowledge for others to benefit from. For all of that, I sincerely thank each one of you, and although there are too many of you to list here, I believe you know who you are. There are two individuals I am compelled to call out, mainly for their direct contribution to this body of work. At the beginning stages of this book, one of my biggest and most enthusiastic supporters unexpectedly passed away. Bobby Halton was by far the biggest cheerleader for this project and a true inspiration for me. We talked at length many times about how to proceed with and shape this body of work, and although I knew I had a monumental task ahead of me, his enthusiasm and endearing encouragement were both infectious and calming, like the strong steady voice of a trusted mentor, and for that I am forever thankful. The second is my friend and colleague, Nick Papa.

Author Leigh H. Shapiro with Captain and author Nick Papa

I first met Nick when he was a teenager visiting my engine company back when I was a company captain. His thirst for knowledge and enthusiasm was unmatched only by my own, and his desire to learn and become a professional firefighter became my personal mission. He would come by the firehouse to hang out, ride along, and soak in every bit of knowledge and experience he could garner. Mentoring is a two-way street, and the efforts imprinted in Nick were equally valuable in the path to my success as well. When it came time for him to test for an eventual career job in the fire service, I again took the mantle and made it my mission to guide and mentor him to a successful outcome. Today, Nick Papa is a rising star in this mission we have chosen, to help others and each other.

Like I always say, I didn't get here by myself; I had a lot of help!

Respectfully,

—Leigh H. Shapiro

Introduction

You passed the promotional exam and now it is hour one of your first shift in that new role—now what?! Where is your head at and how do you plan on effectively functioning in this capacity? You know the material and the expectations, but how do you get there? More importantly, are you prepared and ready for the future? Most officer development programs are primarily focused on the operational and administrative aspects, and very little attention is given to the human element and the dynamics of interacting with different personalities. The operational and documentation aspects of the job, however, are often the most straightforward, whereas navigating the wide range of personnel issues, especially in the firehouse, can be the most nuanced and influential aspect of an officer's job. Because of this, statements like, "Nobody told me that before" or "If I had only known that ahead of time" are frequently heard from new officers being counseled after an incident has occurred. Fire service leadership can be taught, but true knowledge and understanding comes from experience.

This book is designed to enlighten and prepare aspiring and present officers by passing along that wisdom, garnered through actual incidents and events, *ahead of time*. These firsthand accounts serve as the basis for demonstrating how to recognize and process information, both on and off the fireground, to proactively address situations with confidence and resolve. Common oversights and hindrances are identified and addressed to prevent leadership failures reinforced through personal experiences and lessons learned the hard way. Readers will be provided with the skillsets to properly set expectations and drive behaviors. With a roadmap of how to implement leadership and management strategies, readers will possess the direction to effectively perform in any capacity and achieve the desired outcome, setting them up for success.

The material herein is presented as an easier way to do things the right way. Its format reads like a conversation I would have during an actual mentoring opportunity, including some salty language and fire service jargon. There is a clear thread joining information as a mentor and coach would communicate, with learning objectives and relevance, unlike a standard cut-and-dry training manual. This formulates my focus on the personal connection because a mentor must cite examples, anecdotes, and experience for any of the learning to be considered relevant. That's why at times this reads like an autobiographical novel, providing context to facilitate experiential mentoring filling the voids between training and experience. Not everything is learned in a formal structure. Often, much insight is gained at the firehouse kitchen table listening to others speak of their experiences. The structure of the text is presented in bookends format, meaning a scenario is initially presented in the beginning, and carried through at the end of the book.

Upon completion of this book, readers will possess a clear understanding of how to process and utilize prior knowledge and experience in a more effective and efficient way through the employment of critical thinking. By illuminating the nuances of the officer's mindset, readers will enhance their ability to motivate, manage, and most importantly, lead their personnel to implement the interventions that will yield the best possible outcome. Readers will be provided with practical skillsets for mentoring and coaching with an emphasis on personal succession development of both the reader and their personnel. The lessons and strategies shared can be readily applied and allow readers to proactively address situations and avoid the same mistakes made from learning the hard way.

The book begins with the concept and significance of critical thinking and the ability to read both people and environments. Readers are presented with several skillsets to assist in approaching (new) duties and responsibilities. The mental paradigm shift needed for achieving personal succession development and future promotions, goals, and the ability to identify behaviors and mindsets which prompt operational failures and personnel complications are found within the dynamics of mentoring and coaching. Offered are several key takeaway expectations through the insight of parameters and authority of rank, team building, personnel management and expected results to effectively manage and lead simultaneously. Presented is a comparison of the fundamental differences of being "prepared versus being ready."[1] Decision-making is clarified through explanations of several incidents in order to uncomplicate the standard approach to fireground accountability and incident management through the application of incident action templates. A three-point framework is presented to ensure clarity and thoroughness, accountability (from initial alarm to command board),

---

1. Leigh H. Shapiro, "Prepared vs. Ready: Fundamental Differences for Firefighters," *Firefighter Nation* (blog), May 8, 2022, https://www.firefighternation.com/firerescue/prepared-vs-ready-fundamental-differences-for-firefighters/#gref.

incident management (safety measures, active reevaluation, and communication), and simplicity. Anticipating the "Three Common Fireground Failures"[2] and the variables to mitigate them is included. Insight into Cognitive Comprehension Limitations (*National Fire Protection Association [NFPA] 921: Investigations Guide*, Chapter 11 [11.3.1.2])[3] is provided to fully comprehend the human behavior aspect and how heuristic thinking[4] relates to fireground actions. An after-action report is utilized to examine the premise "Encapsulation of the Firefighter: Modern Turnout Gear Equals Modern Problems,"[5] analyzing the introduction of thermal hoods into an urban, career department, as well as the impact and corrective actions taken. A firsthand account and analysis of a line-of-duty death[6] provides operational details, the impact of leadership, and the aftermath both personally and within the organization.

Throughout my time as a company officer, and more recently as a chief officer, I recognized the lack of focus on how to practically handle various situations and incidents, going beyond the generic models and theories provided in basic officer training. My education consisted of merely how to employ standardized, manual-based concepts and methods. Although well-intentioned, they often fell short as their application did not always match the event or the desired outcome and did not account for the variable human element. As I reflected on past incidents, I identified that many of the issues and hardships I endured early on in my career were the result of my approach to the situation and lack of critical thinking skills. While I was competent in the practice of management, I was not equally prepared in the leadership aspect. Because of

---

2. Leigh H. Shapiro, "Recognizing and Overcoming Three Common Fireground Failures," *Fireground Management* (blog), Fire Engineering, July 1, 2018, https://www.fireengineering.com/leadership/fireground-management/recognizing-and-overcoming-three-common-fireground-failures/#gref.

3. Leigh H. Shapiro, "Fire(fighter)-Related Human Behavior: Understanding Cognitive Limitations on the Fireground," *Leadership* (blog), Fire Engineering, February 1, 2022, https://www.fireengineering.com/leadership/firefighter-related-human-behavior-cognitive-limitations/#gref.

4. Ian McCammon, "Heuristic Traps in Recreational Avalanche Accidents: Evidence and Implications," Montana State University, 2004, https://arc.lib.montana.edu/snow-science/objects/issw-2002-244-251.pdf.

5. Leigh H. Shapiro, "Encapsulation of Firefighter Illustrates Need for Critical Thinking Skills," *Leadership* (blog), Fire Engineering, March 1, 2020, https://www.fireengineering.com/leadership/encapsulation-of-firefighter-illustrates-need-for-critical-thinking-skills/#gref.

6. Hartford Fire Department, Firefighter Kevin Bell line-of-duty death while operating at 598 Blue Hills Ave., October 7, 2014.

this deficiency, I was often forced to learn on the fly, or the hard way. After seeking counsel from mentors and formal studies in leadership, motivation, and group dynamics, I have developed a program that provides readers with invaluable insight and guidance to minimize exposure and mitigate risks, increasing effectiveness both in the firehouse and on the fireground. Most of this material is not found in publications and training manuals, but instead harvested from years of tried-and-true experience in real-world situations. The presented material is purposed in the context of personal succession development as well as mentoring and coaching predominately for new or aspiring officers and is in direct reference to my previously published articles.

Without practical education, guidance, and critical thinking skills, those in leadership positions are often thrust into situations where they are charged with making critical decisions without the essential knowledge and skillsets. This lack of awareness is the source of misinterpreting situations, which can have a ripple effect within the organization and the community. Often, excuses regarding the circumstances or being ill prepared are made to justify negative outcomes. This book intends to rectify this issue by providing readers with a practical understanding and appreciation for comprehension gained through experience and engagement with the employment of mentoring and critical thinking skills. With this enlightened mindset, readers will have the capacity to critically evaluate their situation, select the appropriate leadership strategies and tactics, and drive the intended outcome. The objective of this book is to provide readers with a leadership roadmap to enhance their likelihood of success, both on and off the fireground. Because there is no substitute for wisdom, gained only through experience, it is the incumbent responsibility of senior leadership to pass on those lessons to the new or aspiring officers to prepare them for what lies ahead—so they don't have to learn those lessons the hard way. The material is directed toward individuals who function (or aspire to be) in a leadership capacity and those that have a direct impact on personnel, the organization, and the mission by providing practical knowledge beyond training, in order to function in that capacity effectively and efficiently. It is equally beneficial, however, for experienced officers by detailing the modalities of leadership and their strategic and tactical impact to support future reference, guidance, clarification, and enlightenment. The material is delivered in the comprehensive yet simplistic and engaging manner of the mentor or coach, making it suitable for the junior firefighter to the tenured officer. The content is thoroughly supported by an assortment of incorporated visual materials such as on-scene photos, detailed models, citations, and an after-action report from an incident that occurred within the Hartford Fire Department, as well as video screenshots and personal photos for some added humor and most importantly, the stories shared from my experiences.

1

Alarm of Fire!

Tuesday, October 7, 2014, started out as just another workday, or so I thought. First, my regularly assigned driver was on vacation and a detailed firefighter from another firehouse was in his place. On a good day this can spell disaster because of what is expected of our deputy chief's aides, but I knew the replacement well, and I guided him through his expected tasks throughout the day as much as possible when needed. It was a crazy day leading up to the fire. A drowning in a public pool, an emergency medical services (EMS) call for a found suicide, a planning meeting that I was voluntold to go to by the chief of department,[1] and just an overall odd feeling about the day shift, but these things happen and none of us realized something ominous was waiting. By evening, my aide and I sat down for dinner with the other company housed in my quarters, our heavy rescue unit known as Tac-1. The six of us had another stellar firehouse dinner, unrivaled by any of the best restaurants I've ever been to. As we sat at the dinner table relaxing and digesting, having just finished this routine firehouse feast, no one found the strength to get up and start the dishes or begin cleaning the pots and pans and clearing the table. Then the bell hit! The firehouse trip lights alert system came on, instantly illuminating an otherwise dim dining room, and the squawk-box speaker hanging high on the wall near the entranceway began to reverberate with the dispatcher's voice, elevated in tone and with a rapid cadence, which for firefighters is a clear indicator that this call is for real, most likely an actual fire and not a false call, malfunction, or routine alarms activated (fig. 1–1).

As the dispatcher was immediately repeating the assignment over the loudspeakers throughout the firehouse, the crew of the tactical unit was already

---

1. Hartford Fire Department, National Fire Incident Reporting System (NFIRS) 14-280 and District 1 Tour Commander activity log, October 7, 2014.

FIG. 1–1. A time, date, and place that the author will not forget

FIG. 1–2. "Engine 16, Engine 14, Engine 7, Tac-1, Ladder 4, Ladder 3, District 2: 598 Blue Hills Avenue for a structure fire, second floor, we have smoke showing; time of call 18:29!" (Screenshot of video clip courtesy of WTNH News 8.)

Note: NFIRS report 14-02800045, "Incident Dispatch, 598 Blue Hills Ave.," October 7, 2014.

headed down the stairwell onto the apparatus floor (fig. 1–2). The crammed and weighty Ferrara rescue truck came to life and roared out of the firehouse through a cloud of soot with both its diesel engine and the Federal Q siren screaming with unyielding urgency! The district chief unit which I am assigned to does not respond to a call in the north end of the city where this incident is located unless it's requested, the other district chief is unavailable, or it's a confirmed working fire,[2] so I got up and ran to my office to grab the desktop

---

2. Hartford Fire Department, *Administrative Manual—Department Directives*, 2015.

portable radio to listen to further communications, and also unashamedly because I'm a fire buff at heart. The address where the fire was reported was only two houses to the south of Engine 16's quarters, a single-engine firehouse on the same side of the street. They arrived first and confirmed a working fire in their first-due size-up radio report. As soon as I heard this radio transmission, both myself and my driver were already headed down the stairs to our chief's sports utility vehicle to respond to the fire, mandated by our department directives which indicated I would be the designated safety officer at this fire and report to the incident commander (IC).

While en route, I listened to multiple radio transmissions from companies operating on scene, some of which were garbled with static and unintelligible. The transmissions were the typical radio traffic which occurs at a working fire: companies getting direction, companies asking to charge various handline and supply lines, and the normal notifications that need to be made during this type of event (power company, gas company, fire marshal's office, etc.). Once we arrived, I put my gear on and reported to the IC. That's when this incident began to go sideways. I walked down the driveway on the B side of the building to the rear and observed a bedroom window with heavy fire emanating from it. Then I turned directly around and carefully observed the exposure building immediately next door approximately 20 feet away from the fire building, which was another wood frame house similar to the one on fire. The fire had not communicated to the exposure yet, but it was just a matter of a minute or two until it did, so I rushed back up the driveway to the IC who was now standing at the entrance of the driveway, and I asked him forcefully "Where the f@#k is 16's, why aren't they making the push?"[3] This was my thinking because I had responded from my downtown quarters on the report of the working fire, and it took several minutes for us to travel to the north end address, park, put on our gear, and walk over to the command post. About 15 minutes had already passed when I asked the question about the first-due engine, housed literally two buildings away from this working fire, yet not having a drop of water flowing. This to me was very irregular. Something was wrong. Just as I had finished my sentence to the IC, our attention was immediately drawn to the front of the structure. The second-floor window had just been smashed out and leaning out of this window was a firefighter with no helmet, hood, or mask on. Before any of us could react, the firefighter tumbled headfirst out this window and landed on his back onto a large bush just below, which most likely broke his fall. I spun right back around to the IC's bewildered gaze and yelled "Get them the f@#k out of there, everybody out; dump the building!!!!" (fig. 1–3).

---

3. Connecticut State Police, Case Number CFS1400630622, October 9, 2014.

FIG. 1–3. A command to order everyone out of the building (screenshot of video clip courtesy of WTNH News 8)

Literally all hell just broke loose: the IC was screaming on the radio giving the evacuation order, other companies out in front were running over to give aid to their injured brother, and another company was throwing a ground ladder to the window we all observed the firefighter come out of moments earlier, thinking others were surely to follow.[4]

That firefighter was immediately moved to an awaiting ambulance stretcher and whisked off to a nearby trauma center right down the street. I turned to my aide and ordered him to request a second alarm assignment at this incident from the dispatcher. Meanwhile, as companies were exiting and forming out in front of the fire building by the command post, the IC began his radio personnel accountability report (PAR) by asking each company on scene if they had PAR, meaning if all their crew personnel were accounted for. After a minute or so, the lieutenant of the first-due engine company approached the command post where myself, the IC, and both aides were standing and said, "I can't find Kevin Bell!" I immediately barked back at him "What do you mean you can't find him; where did you leave him?!"[5] He said he was last seen on the second floor. The IC immediately activated the rapid intervention team (RIT) that was standing by on the front lawn. They grabbed a charged handline that was lying in front of them unattended and barreled through the front door up to the second floor to search for their missing brother. Things were not turning out as expected for this simple bedroom fire. A minute or two passed during this intense, nail-biting search. The atmosphere was thick with tension, and the crisp autumn evening air was overloaded with sirens from responding companies and ambulances. Just then I heard someone from inside the

---

4. Connecticut State Police, CFS1400630622.
5. Connecticut State Police, CFS1400630622.

structure yell "We found him—we got him!" Kevin was found unconscious, tangled in the legs of an overturned table, just inside the right-side front living room at the top of the stairs of this two-story duplex.[6]

The RIT company along with a ladder company aggressively worked to disentangle him, which felt like it took longer than it actually did, then swiftly brought him down the stairs to the outside front door walkway where I was standing. When they finally got him outside, he was literally laid at my awaiting feet, and multiple firefighters and EMS personnel pounced on him to get his gear off and assess his condition. His water-soaked gear was intact, and he still had his self-contained breathing apparatus mask affixed to his face; however, Kevin was obviously unconscious. This was a sight no one would ever forget. Someone from EMS rolled up a stretcher to where we were, and Kevin was briskly tossed onto it and forcefully loaded into an awaiting ambulance. Off they went at a high rate of speed, sirens blaring, with police cars both in front and behind the ambulance. I quickly turned back to the IC and stated to him "We still have a fire to put out!"[7] We both proceeded to put the pieces back together from the resources we had remaining to subdue the remainder of fire.

During the remainder of suppression activities including salvage and overhaul, I was kept informed via cellphone of Kevin's status from the Employee Assistance Program (EAP) coordinator who was at the emergency room with Kevin's family. He was a detailed firefighter to the office of EAP, one whom I knew well and trusted, and always spoke with a sense of gravity and humility. Periodically, while waiting in the hallway just outside the trauma room where the emergency room (ER) staff were desperately working to save Kevin's life, he would ask one of the doctors or nurses what was happening, and then would call me immediately and relay that info directly. No one else at the scene of the fire had this information except me, and I kept it to myself and the IC, because I did not want to alarm any of the firefighters. This was a highly volatile, tense situation and injecting this type of ultra-sensitive information into the crew members at the fire scene would be counterproductive and an unwanted distraction. The telephone reports I was receiving from the ER were that Kevin had regained and then lost his heartbeat several times over the course of the two hours since he was transported. For several of these phone updates, the EAP coordinator stated, "It's touch and go!" Hearing that phrase now brings back all the emotion I was feeling as when I was on scene, standing on the

---

6. National Institute of Occupational Safety and Health (NIOSH), "Career Fire Fighter Dies from an Out-of-air Emergency in an Apartment Building Fire—Connecticut," Fire Fighter Investigation and Prevention Program F2014-19, January 24, 2017, https://www.cdc.gov/niosh/fire/pdfs/face201419.pdf.

7. Connecticut State Police, CFS1400630622.

sidewalk. Having had years of experience in EMS, I inherently knew this was not good. The IC and I cautiously supervised the remaining elements of this fire incident and slowly initiated the companies to pick up their tools, equipment, and hoselines and place them back on their respective apparatus. By then the Hartford Fire Marshal's office inspectors were on the scene, along with several Hartford Police officers, detectives, and Hartford's police chief and assistant chiefs were there as well. This was a very serious call, and it was all hands on deck. Firefighters, support personnel, fire photographers, news media, EMS personnel, and utility company personnel were all milling around at various locations in front of the fire building when the final pieces of hose and equipment were placed back on the apparatus. Firefighters were grouped together quietly talking about what had transpired. Many were in disbelief and shock. As for me, I was already stunned, having just dealt with the worst thing ever in my career. Things were about to get a whole lot worse.

Case Study

After most of the firefighters on scene at this fire scenario either witnessed what transpired or had directly participated in the search, rescue, and removal of the downed firefighters, they clearly have lost their initiative and are now completely distracted by what transpired. This has not only been a horrific turn of events, but two of their brother firefighters are now at the hospital fighting for their lives. Nobody is in the mood, so to speak, to put a fire out; they are entirely consumed with emotion, grief, and anxiety, and this has a direct effect on the remainder of the incident.

Questions

1. What steps would you take as a company officer to reestablish focus on the tasks ordered by the IC, after either participating in or witnessing these emotional events at this incident?
2. What steps would you as the IC take to reestablish an effective command and wrangle the troops to complete their assigned tasks?
3. As IC, what other resources would you summon to assist you in completing the safe mitigation of this incident?

2
Mentoring and Coaching the Fire Officer: Pearls of Wisdom

You just got your officer's badge pinned onto your uniform by either a loved one or a respected peer. This is one of the happiest days of your fire service path you chose, either as a professional paid career or a dedicated volunteer. Now what?! You have endlessly prepared for this day and this moment, and it's your time to set the world on fire with your newfound authority and responsibilities. You've earned it, right? Not so fast, kid! There's more to it than just a promotion, a badge, and some memorized knowledge. For each of my three promotions, when the badge was pinned into my uniform, the weight, gravity, and seriousness of these newfound responsibilities increased each time, culminating by ultimately receiving my gold chief's badge. The weight of the entire department was on my shoulders, as well as the expectations of both those who came before me and those who would follow in their professional paths. This is serious business. We are the people that show up to someone's worst day, whether it's a fire, emergency medical services (EMS) call, or some other type of emergency. We show up and do what we are trained to do, without question, hesitation, or timidity! To do this job the right way, this chosen profession of saving lives, stabilizing incidents, and property conservation, there needs to be constant training and education. But what do you do when your foundation of training and education is limited? You go outside the box, the comfort zone, and at times, your department, and seek out knowledge wherever it may be. For me, beyond books, classes, videos, and conferences was the more intimate and personally defining engagement of mentoring. This is where I learned the true craft of *firemanship*, bravery, respect, commitment, dependability, dedication, and servant leadership. There is, however, a clear delineation between mentoring and coaching. Coaching is the act of receiving instruction and advice to further a goal, whereas mentoring is an experienced, trusted, wise, and often senior individual influencing and counseling someone else within their career and often beyond. Let's dig into what they don't teach you in school.

Ask any seasoned or salty firefighter and they will kindly inform you that you need to take this job seriously while simultaneously not taking yourself too seriously (fig. 2–1). The responsibilities of being a firefighter is serious business. We are tasked with doing our very best with our training and education to mitigate an incident or event which may be someone's worst day ever, or some tragic, horrible, life-altering event. Throughout my experience in this field, dark humor, locker room banter, and at times, immature behavior (all in private of course) often provided the necessary mental relief needed to stave off going down the dark hole of drugs, alcohol, abusive habits, and all the other associated coping mechanisms human beings utilize to deal (or not deal) with their issues. Many of these problems are unexpectedly brought on by the daily stressors of the job by its very nature. And while expounding on mechanisms

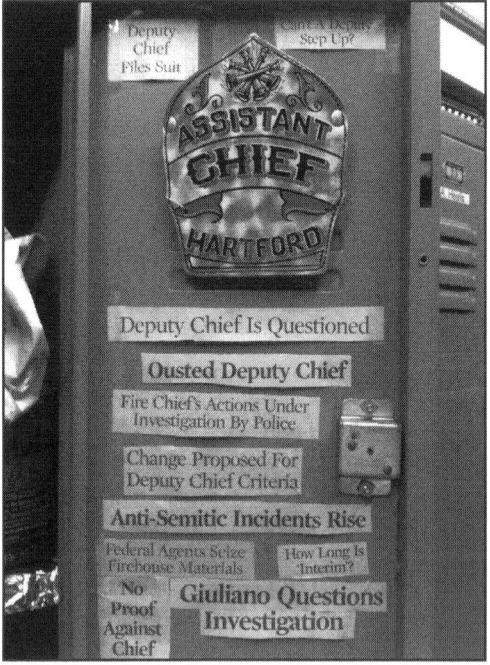

FIG. 2–1. Taped to the inside of my food locker in the kitchen/dining room of my firehouse is a series of random, coincidental newspaper headlines collected over the years that provided pointed comic relief. The running gag in the firehouse was that somehow, we were always in trouble. Even if we hypothetically rescued a cat out of a tree, we joked we would be sued for animal abuse by the owner because being rescued was not what the cat wanted! Anthony Giuliano Jr. was my assigned deputy chief (DC) aide, and the headline bearing his name was yet another convenient and humorous coincidence. The assistant chief helmet shield was issued to me during my time as interim assistant chief (AC).

of enlightenment, let me also say that I have always maintained that it is better to get it right than to be right, meaning you don't have to always have the last word or be the smartest person in the room. It's better to simply get it right, meaning whatever you are talking about or doing, do it properly and the right way (safely). It's not a competition of the witty! I don't think there is a way other than by physical force that you can teach someone humility and empathy, which are the foundational tenets of servant leadership, which means you serve a greater cause than your own. For me, this has been the most effective way to employ being a leader and getting your crew to buy in to what you are selling, which is leadership and management. You garner your knowledge from wherever you can, but to apply it you need training. Think about the difference between someone who read a book and now believes they are a professional firefighter, and the recruit or newly appointed officer who received formal training, constantly practices, and executes the knowledge in a safe, practical manner. Furthermore, it's okay to give credit and acknowledgment as well, instead of your only motivation being receiving validation, attention, and recognition. Many a probie have been admonished to keep their mouth shut and their ears open. The strong, silent type may appeal to many and indicate a distinct sign and symbol of leadership and maturity. Conversely, however, it can be a glaring indication of inferior leadership using silence to shield from being exposed as a fraud. Be mindful.

New Duties and Responsibilities

Now that you're officially a fire officer or an acting fire officer—depending on where you are in your career path—you will have a whole host of new duties and responsibilities, some of which you're just finding out for the first time. Many of these were spelled out in your job description when you applied for the recruit firefighter job or the officer's promotion. Education is what you know, but training is what you do. Knowledge is great to acquire, but if you can't put that into action, what good is any of it?! As a firefighter, and especially as an officer, you now represent your department, your city or town, and your jurisdiction. This means that when you respond to calls for service, interact with the public, and perform your official duties and tasks, you are representing. You are the representative of your department, its core values, its mission, and vision statements. If you don't know what each one is, you should probably investigate and find out, and if they don't exist, start asking why. Every department needs these esteemed declarations.

In the fire service, departments and agencies are formulated similarly to military structure and operations in that they are scalar in format. The scalar organization has leadership at the top as the chief of the department, and underneath that person are various ranks leading all the way down to the firefighter, most often shaped like a pyramid. You, too, have your specific role within that scalar organization. Every rank must answer to another rank, and for some, this can be a tricky task to comply with because it limits the latitude for independent judgement and stifles freelancing.

In the absence of the fire chief, the fire officer assumes the responsibilities of the chief of department, in that you represent the department;[1] no, you don't hold the same authority and responsibility, and you certainly are not getting paid the same, if at all, but if you are the only officer in uniform at a scene or event, then you carry the burden of representation. If you are at an event in which your department was represented, such as a training session, a social event, community organizational event; at an incident where you are the (only) officer; or if you are not the only officer there but in fact may be the highest-ranking, you represent your department as the officer. I recall several times in my career being the only chief officer among a large contingent of Hartford firefighters attending line-of-duty death funerals, and as the ranking officer, I represented my department in an official capacity, not by being asked to or forced, but simply by default. Other times, when attending many neighborhood community meetings in my fire district as a company lieutenant or captain, I represented the department (as the default chief of department) at these meetings. I may not have known the specific agenda of the chief, but I didn't need to know. My presence represented my department and its core mission, values, and vision statement. If asked a specific question that I didn't know the answer to, I simply replied, "Please give me your contact info, and I'll get back to you as soon as I have an answer." I didn't cower, ignore people, or shy away from engagement. I simply worked with what I had. If the chief really needed to be in attendance, they would be.

Many firefighters find themselves in hot water because they engage in unflattering or unprofessional conduct while off duty, which is your prerogative to do, as long as you aren't wearing a department t-shirt or hat or have on your patch-laden job coat for the whole world to identify you by. Even when you are not on duty, if you project the image and identity of a firefighter, you are representing your department and the chief. Recently, I was with a local department when their chief told me that one of his firefighters was seen on a Facebook clip engaging in unprofessional negative behavior at a political rally the day

---

1. Hartford Fire Department, *Administrative Manual—Department Directives*, 2015.

before while wearing a department t-shirt. Per department standards and rules, the firefighter was dismissed from the department by the next business day. No joke: you're out there, you're representing, period.

Like an old dog, even the most tenured, senior personnel within any department still learn new things and stay current. You need to keep your skills sharp and stay involved in the training and learning aspect of your job, especially when you are an officer, because now your personnel are looking to you for answers and direction. If you think you know everything and claim to be some type of an expert, you'd be wrong. The plain truth of this matter is you always need to be a student of the fire service. Why would you not want to learn as much as possible about a chosen career path that may end up killing you or those with you? The dynamics of this business dictate that things constantly change, mostly for the better, but bad things still happen within the fire service. After retiring from Hartford, I engaged in the process of obtaining my Connecticut Fire Marshal certification believing it would enhance my future job prospects. After over 30 years in the business, I sat in class each day for 6 months literally mumbling out loud, "Holy crap—I didn't know that!" Even after my fire service career in Hartford sunsetted, I was still learning new information and gaining enlightenment in our fire service universe.

Firefighters and officers need to maintain situational awareness wherever they are, whether it's at a function, in a training event, and obviously while operating at an incident. Situational awareness is paramount to embracing your role, especially as the officer. If you don't know where you are or who you're with, or if you don't know the situation you're dealing with, you are working blindly. As an officer, when people come to you with questions or ask you for guidance, the only meaningful way you can answer that is with situational awareness. We all can recall working with firefighters and officers who may have been less than informed or just plain detached from their own reality. Not a good characteristic to be associated with. You really need to know both where you are and what's happening around you. Why, you ask? Because it could save your life or those with you. Study the map, study your district, and pay attention to the buildings you enter, especially the egress and fire systems. Walk around inside and outside of these structures. Ask questions, be nosey, take initiative, and do your job! At incidents, fight the path-of-least-resistance urge to get tunnel vision. Pay attention to the radio transmissions. Observe and learn. Just as when riding a motorcycle, keep your head on a swivel, because you never know what may be right around the corner. In today's crazy environment, individuals in uniform are often seen as a threat rather than a helping entity. Be ready for unanticipated pushback, sometimes in the form of violent behavior. There is no duller a knife than that which refuses to be sharpened!

Bear in mind you are a sworn professional. Most if not all professional departments and agencies require that you be sworn in and take an oath of allegiance to the jurisdictional charter or whatever they may call it. By definition, the term *professional* means someone who performs the respective duties and responsibilities of their job to the best of their ability, irrespective of their personal sentiments.[2] That means when you raised your hand, took that oath, and were sworn in, you established an obligation and commitment to fulfill your duties the best way you can regardless of how you feel about the situation, who you're dealing with, or what shortcomings or obstacles you may encounter. And, as a fire officer at the rank of captain or above, in many departments, you are now a leader of leaders, meaning you may oversee other leaders and they, along with firefighters, are now looking to you for answers, guidance, and leadership. Let that sink in for a minute or two: the leader of leaders. Those are some big shoes to fill, but like most firefighters, we rise to the challenge and excel at our mission. Overseeing other leaders comes with a whole new bag of tricks you need to be familiar with. Always being on point is an understatement. You now must demonstrate the knowledge, skills, and abilities to not only do your own job, but also the job of those under you in this scalar organization as well. These leaders will surely call you out if they think you are wrong, on the wrong side of an issue, or just plain suck at what you do. One of my former colleagues recently received his promotion to district chief. Now he oversees a group of fire companies and officers on a particular shift. I texted him a heartfelt congratulations, and in the same message said, "It's for real now!"—meaning he is now in charge of leaders and carries a huge responsibility. The training wheels are off; you need results, not excuses.

As a fire officer, you have a broader role, authority, and responsibility when engaging your daily activities and your overall job duties. You now have a company to both lead and manage, a firehouse with multiple shifts and crews you may oversee, and you could be assigned to a shift where you often must provide guidance and direction. As a career professional firefighter, you must abide by your collective bargaining agreement (CBA), also known as your contract, if your department has one. Many departments throughout the country do not, either because they are a right-to-work state or because they are a volunteer organization. Certainly, you now have a broader role and responsibility not only within your department, but to your city, town, or jurisdiction as well. It's no longer you just riding the back step assigned to the hydrant, pipe (nozzle), roof, or irons riding positions, as we say in Hartford. Now you have serious

---

2. "Professional," Dictionary.com, https://www.dictionary.com/browse/professional.

responsibilities beyond the scope and duties of a firefighter. So how do you approach your new duties? Where do you start? For me, after our first year on the job after graduating from the fire academy, which was also a full year of probation, we were expected to train in house to prepare to be in charge or to sit in the front seat when there was no company officer on duty in our assigned crew. We rotated on a fair and equal basis with each member of the company, but when it was your turn, you were up! Beginning very early in my career, I got a lot of trigger time being in charge of a fire company and all the experiences that went along with that. After 5 years on the line, we were eligible to take the lieutenant's exam, and if you were paying attention for the last four years, you had a pretty good feel for the job and what it took to do it properly (fig 2–2).

For the newly promoted fire officer, there are several principles that help maintain your focus, direction, and purpose, which I found very helpful. Unfortunately, both new and seasoned officers sometimes need to be reminded that they are still a firefighter first, meaning regardless of your stature, rank, or authority, you are a firefighter! Your core mission has not changed, only your rank and duties, when you receive a promotion or other positions of responsibility. This means relying on back-to-basics at times. For me, it meant no matter what I was dealing with when I was a lieutenant, captain, and deputy chief, I maintained that my true core was firefighter first! Rank, titles, and authority can be transient, but when you raise your hand and are sworn in as a firefighter, that is when you agree to commit to the mission of the fire service, on and off duty, whether ranking officer or simple firefighter. You chose this career path, with all its highs and lows, and there are untold lows in this job: nights, holidays, weekends away from family, hot and cold working environments, uncooperative public, administration interference, and so on. However, don't forget where you started and where you came from, as well as those that

FIG. 2–2. Pictured is the 2017 recruit class which my youngest son, Quentin, was a graduate of (middle row, obscured). Now all of these recruits have broader roles, authorities, and responsibilities, which are multiplied and amplified when someone is promoted to fire officer.

came before you and will come after you. For the officer, hopefully your promotional opportunity was equal for everyone who took the exam. Obviously, you scored high enough to reach the promotional point. Approach your duties with humility, dignity, and respect for not only yourself but for others as well. It is also beneficial to simply say please and thank you within your interactions with people. There is a time and place for the autocratic dictator in the fire service, but not as often as you may think. Intertwined among all the other responsibilities as an officer, one of your not-so-obvious primary goals and constant objectives is to prevent negligence, meaning to make sure things are done properly, effectively, and efficiently. Follow your rules, regulations, and department operating guidelines as well as your CBA if you have one. Your function is to perform a job for which you were promoted to do, and there are serious ramifications if it's not carried out thoroughly and successfully. It is just as easy to get into hot water by not doing something as it is when you are doing something. I once asked my department chief why we could not simply do something that seemed obvious and easy, and he responded, "Your job is to prevent negligence!," which I did not fully understand at the time. Even simple decisions can have major implications, and as the officer, you're responsible for preventing those negative results. If they do occur, and they often do, you must answer for them (figs. 2–3a and b).

FIGS. 2–3a and b. This June 1999 general-alarm fire in Hartford was about as crazy as one incident can get. Flying embers were responsible for numerous secondary multiple-alarm structure fires throughout the neighborhood. Mutual aid companies assisted but quickly found themselves overwhelmed and overrun. My company, Engine 16, dropped our entire supply line bed, took up a position at the corner of the structure, and fed a tower ladder and a deck gun simultaneously. Throughout the duration of this chaotic incident, the leadership capacity and fortitude of all the fire officers was continuously tested. See a young Lieutenant Shapiro atop the Grumman Wildcat! (Photos courtesy of Alan Chaniewski and the Hartford Fire Department [HFD])

Critical Thinking for Fire Officer Development

One tried-and-true technique employed throughout my years of service is the engagement of critical thinking and the skillsets required for such. I was more effective as a firefighter and leader when fully understanding what and why the situation was being presented in its form. The concept of critical thinking has been around for years, but only more recently applied to our line of work. The fire marshal world has been utilizing this method for many years, most notably as spelled out by *National Fire Protection Association (NFPA) 921: Guide for Fire and Explosion Investigations* (fig. 2–4).[3]

NFPA 921 states each investigation shall be conducted in a methodical order from one to seven:[4]

1. Recognize the need for an investigation.
2. Define the problem.
3. Collect data.
4. Analyze the data.
5. Develop a hypothesis by ruling in knowledge.
6. Test the hypotheses by ruling out evidence.
7. Select the final hypothesis.

For us firefighters, it presents clear, rational, open-minded, disciplined thinking informed by evidence. It sounds like a mouthful, but instead of fumbling through a situation or scenario, applying critical thinking assists you in prevailing. It allows you to identify links between ideas based on importance and relevance. Then you can recognize and build an argument and identify inconsistencies and errors (fig. 2–5). It provides the ability to be consistent and systematic for you to make the justification needed for your decision process.

So, you ask, how does this all relate to the fire service? Although the fire academy provides the necessary skillsets to become a firefighter, it is wholly based on the *act more, think less* process. Instructors instill algorithms and muscle memory into new recruits so they may function efficiently at incidents without having to relearn and rethink the fundamentals. This is great but there is no emphasis on processing information. Especially with fire officers, there is no substitute for experience, and replacing muscle memory with critical

---

3. *NFPA 921: Guide for Fire and Explosion Investigations* (Quincy, MA: NFPA, 2021).
4. NFPA 921.

thinking is a tried-and-true methodology. It applies clarity of thought, not simply decisions based on attitude. It allows you to steer clear of groupthink and compels you to engage emotional intelligence (fig. 2–6).

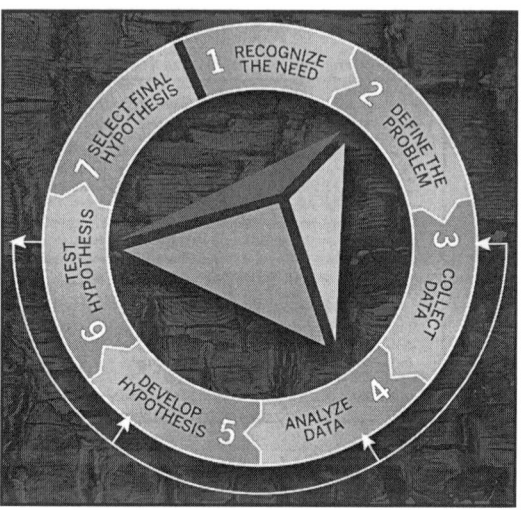

FIG. 2–4. The scientific method for fire investigations is deemed critically vital to properly investigating fires and is demonstrated in this graphic which is on the cover of every edition of the NFPA 921 guidebook. It is based on the concept of critical thinking.

FIG. 2–5. Critical thinking was clearly absent from the decision to chew down this tree. (Photo courtesy of Quora.com)

Critical thinking is the process of objective analysis and evaluation to formulate a judgement.[5] Basically, it provides context and perspective in the decision-making process. For officers, it facilitates the ability to quickly

---

5. "Critical Thinking," Dictionary.com, https://www.dictionary.com/browse/critical-thinking.

FIG. 2–6. Critical thinking facilitates context and perspective for the individual. The firefighter assigned to connect the supply line to this hydrant wasn't thinking clearly when the decision was made to kink it around a utility pole and wrap it around an old firebox stand. (Photo courtesy of HFD)

analyze a situation, often when processing a large amount of information. That information at times can be incomplete or just plain wrong, so it provides the necessary proficiency in knowledge, skills, and abilities (KSAs).[6] In certain situations, a wholesale loss of institutional knowledge compels firefighters and officers to engage in critical thinking more often than expected. When I retired in June of 2016, myself and approximately 250 fellow firefighters and fire officers walked out the door within the span of about 6 months, leaving the administration and those remaining firefighters to figure it out for themselves. An idiotic and selfish decision on behalf of the mayor and the fire chief allowed this to happen, but they were only interested in the budget, not the impact of the loss. Since then, time and grade standards have been modified to allow what's referred to as the *not-ready-for-prime-time players* to be quickly promoted up the ranks without any real experience, preparation, or training.

Mentoring Works

What creates the synergy among an effective and efficient crew of firefighters on any given apparatus, in a firehouse, or on a particular shift? You know it when you don't see it, right? We've all grumbled beneath our frustrated breath at some point over a not very effective crew. So, what exactly is it—the right combination of people, the right time, or the right motivation? Could it be sheer luck, or is it by design or engineering? Why are some crews great at what they do, and others simply show up for work or are merely thought of as the help? Are we each viewed as individuals or are we cohorts with an enduring bond or creed, or maybe professional colleagues striving for the best that we can

---

6. NFPA. Job Performance Requirements (JPRs) and Knowledge, Skills, and Abilities (KSAs), nfpa.org.

deliver? What causes firefighters working closely together to go above and beyond what is expected, while others put just as much effort into being disengaged and uninformed as possible with expected minimal results? What is the mortar between these bricks made of, so to speak? The answer is in how mentoring influences and guides firefighters and officers alike. This positive input, usually combined with institutional knowledge, experience, and enthusiastic magnetism, ultimately empowers ambition, energizes morale, and provides direction and guidance needed to fill the gaps in training and experience. This almost always produces an enhanced quality result: a better firefighter. Need proof? How many times have we seen dysfunctional crews, firehouses, or shifts turned around by effective leadership? Providing strong guidance with clear expectations, setting the tone, administering inclusive authority while simultaneously maintaining responsibility for the outcome, actually knowing your job instead of repeatedly professing that you know your job, unwavering and steadfast humility, and just plain respecting everyone for who they are is what I found most effective during my tenure and beyond. And yes, sometimes the squeaky wheel crew member doesn't always get the oil; they get replaced! But to merely attribute this solely to leadership then raises the question, what drives that leadership? What is it that makes an effective leader? If I buy a hammer at a hardware store, am I now a carpenter or, worse, a professional? No; thankfully there's more to it.

FIG. 2–7. A skilled, experienced 16-year department veteran and lieutenant of its heavy rescue unit mentors students on the most effective and efficient way to conduct a rapid search inside a fire building. Such mentoring is invaluable because it communicates relevant coaching beyond the student's training and experience.

Mentoring is a peer-to-peer, two-way street and moves in any direction, even to somebody outside your department (fig. 2–7). It can flow upward to higher ranks and flow downward to the probies and new members. It can be delivered individually or in group settings.

However, merely citing and applying the term *leadership* is too broad and really doesn't define the core transaction of learning. Coaching and guidance from a mentor is not simply the act of receiving information and direction when needed, which can be situational and fleeting. Mentoring can often show you, beyond telling you, how to carry yourself and conduct your business, both professionally and personally, especially when interacting with fellow firefighters, the administration, and your core customer base, the public. Sometimes, mentoring can even be more effective at mitigating screw-ups and bad actors than progressive discipline can. Basically, it can serve as the needed direction in someone's career and even in their life, whether in the immediate or long-term, and inspire someone to emulate characteristics and define themself. It can also sustain hope when you may need it most. Years ago, when I tested and ranked too low on the deputy chief's list to be promoted, several mentors recognized my self-inflicted anguish and carefully proceeded to reaffirm my value, restate my purpose, and build me back up mentally to where I needed to be and to where they expected me to remain. They propelled me on the course of resolve and preparation for the next scheduled exam (3 years later), for which I ultimately scored number one on the list and got promoted to deputy chief. Their guidance instilled in me the confidence, fortitude, and perseverance I needed to take full advantage of the next few years and build my level of training, education, and experience, which was vital to achieve my goal.

There are many who profess KSAs, but for these to be effective for you personally, you need to find that which you are comfortable with and what works for you. Many people will provide you with mentoring, but you need to decide if you want to engage that style and message. You need to figure out what works best for you and not merely follow someone blindly because they have a title or KSAs. Several individuals I considered mentors throughout my career have shown me how to do it, and how not to do it—both the good examples and the bad examples (fig. 2–8)! I learned early on not to judge by appearances—everyone you interact with has some takeaway knowledge for you. You just need to recognize and identify it, then decide what to do with it.

The Red Car

Although many names come to mind, I distinctly remember what each mentor taught me, both the big concepts and the quiet little details. When I was a captain, I was assigned my first shift as acting deputy chief in the district sports utility vehicle, known to Hartford Fire as the Red Car. I was working directly

FIG. 2–8. At this two-alarm house fire, I as the AC mentored the junior chief officer, who was the incident commander, assisting with thoroughness, organization, and peripheral responsibilities (other incidents occurring elsewhere in the city). (Photo courtesy of Pat Dooley)

with a chief's aide (affectionately called Red Car drivers) and would respond to calls within a given assignment and act as the chief officer running the operation. Although I was prepared to rise to the challenge with my skill sets, it was the Red Car drivers who taught me and everyone who served in that position the job of DC. On my first day, I walked into the district chief's office, fully expecting to be shown around, to receive guidance and direction, and to ease into my new duties and responsibilities. Instead, the chief officer I was relieving simply said, "Sit here!" and pointed to the desk chair, and then said, "When the bell hits, get in the car!" and walked out of the office. That's it—that's all I got from him! I stood in the office for a brief minute when, suddenly, the firehouse alert system sprang to life reporting a neighborhood church on fire down the street, of all things to start my new responsibilities! After the dispatcher gave out the assignment, the office phone rang where I was standing. Dumbfounded and trying to keep my emotions in check, I answered it quickly. It was the chief officer I had just relieved, who had gone to the kitchen to get some coffee and was calling me from just across the apparatus bay floor. When he spoke, I could hear him both over the phone and from the kitchen, because he was laughing loudly when he said, "Funny thing about those church fires—you can never really put them out!" and then slammed the phone down! Just then, the Red Car driver stepped into the office to make sure I knew we had a call, and that's when I realized I would be alright. For the next 10 years, I was fortunate to have sought out and gravitated to excellent leaders and mentors to learn my craft as a chief officer, so when I was eventually promoted into that position, I was prepared but, more importantly, ready to effectively fulfill my duties and responsibilities. That church fire call turned out to be a boiler malfunction—no biggie!

When I was a newly minted lieutenant assigned to a busy engine company, I had acquired the KSAs to perform my new duties, but it was the very seasoned apparatus operator with whom I was assigned who really spent the time showing me the ways of the job. He was indispensable, an invaluable and trusted mentor during my time with him. The knowledge I learned from him helped guide me for the remainder of my career.

What Makes the Difference?

What it is about a particular individual you've interacted with, perhaps a long time ago or maybe recently, that has such staying power that it gets stuck in your head and even creeps into your bones? Most firefighters aren't looking for a friend; they're looking for guidance and validation. They are seeking confirmation that their input and effort is important, necessary, valuable, and appreciated for being part of our overall mission in the fire service. We all have opportunities at some point: attend school, participate in training, go to this call or that incident, get involved with certain people or groups, and so forth. To me, all that is good, but the real rub here is not what you have done, where you have been, or with whom you are seen; those are simply opportunities. The real power comes from what you have learned. What did you learn from those opportunities, mentors, and coaches, both good and bad, that compels you to think and teaches you to be a better person, a better firefighter, and a better leader? What has changed inside your mind for you to understand that one behavior is far better and more effective than another, or that amplifies the gravity of not only knowledge but what to do with it as well? Achieving your goals, both personal and professional, derives from believing in the messenger, the message, and the purpose when receiving guidance and mentoring. Wisdom is sought and valued, and trust verifies and validates that knowledge. During those times when I was preparing for promotion, I would seek out individuals

I trusted based on my interaction with them, their performance as firefighters or officers, and the whether I believed what they would teach me was relevant, worthy, and true. What I learned from all those individuals in aggregate is what I embrace and depend on when providing mentorship to those seeking experienced and prudent advice, clear-minded judgement, guidance, and wisdom learned from my own career. I have made plenty of mistakes but have learned something from each one. Sometimes, I had to repeat those missteps to appreciate the correction, but if I had a dollar for each blunder...

There are many academic manuals, essays, lectures, and coursework that eloquently describe the various types and styles of leadership to dissect their formulas and understand the sum of their parts, but firefighting crews and individual firefighters will more likely counter with a name, not a theory! They tell how a particular leader, mentor, or individual they knew or worked with made them feel, unlocked their understanding, motivated them to achieve, and empowered them to believe in themselves. People come and go in the fire service, but the transformation and measured impact garnered from outstanding leaders and mentors alike linger like a certain smell, an old song, or a tried-and-true recipe. That's the real value of mentoring.[7]

Plenty has been written on the topic of leadership detailing the variety of styles and their applications. Of all the types, one has always intrigued me because of its tangible results and lasting impact: transformational leadership. Defined as the initiation and resultant change within an individual, a group, or an entire organization, it achieves the desired outcome by illuminating, inspiring, and empowering. This approach is how KSAs take shape and are readily measured in its effectiveness.

Transformational leadership is a highly effective style of personal leadership because of what it ultimately achieves and how it is accomplished. As a leader, I often asked myself, "What does it take to change someone for the better if, in fact, it can be done at all?!" I believe it cannot come from an authoritative standpoint, much like a transactional leader or manager who is narrowly focused on immediate tasks and singular goals. It also cannot be attributed simply to a personality trait or character element, a regurgitation of a leadership or management concept or core value statement, or, as in some cases, forced or coerced. And to only identify training as the explanation is not enough. If I teach someone how to boil a hot dog in a pot of water, does that make them a chef? No; this goes far beyond that. Transformational leaders possess the ability to positively influence others by motivating them

---

7. Leigh H. Shapiro, "The Real Value of Mentoring in the Fire Service," *Leadership* (blog), Fire Engineering, August 1, 2023, https://www.fireengineering.com/leadership/the-real-value-of-fire-service-mentoring/#gref.

intrinsically. They define expectations and provide the support to achieve a goal beyond what was thought to be possible. These leaders embrace the core mission of the agency, the value of their contributions to it, and the understanding of the requisite big picture of getting all the other fishes to swim in the same direction to advance the mission. The key to triggering this change is through empowerment. The concept of experience has produced the most reliable and impactful results. Empowerment often comes from delegating or granting permission to a crewmember or company by way of an assignment, such as when I state, "I need this done; please go do it!" Other times, it's out of necessity, such as when you pull up to a working fire and you're the only officer on scene—so it's time to step up and run the show! To stimulate empowerment from someone who already has the authority to make decisions, convey the responsibility element (such as giving an individual or crew an order or task to complete) and let them figure out how to make it happen (with your only interest being the result). You can see in their eyes the level of commitment and the resultant satisfaction when they are afforded an opportunity to excel and the latitude to self-regulate without the fear of being humiliated. Instead, they are coached only *if* they fail.

During my time as tour commander, a lieutenant on my shift was frequently the first-arriving officer at working fires (fig. 2–9). He would initiate command, immediately call for more resources (when he deemed necessary), and then proceed to apologize profusely to me when I arrived on scene for doing exactly what was expected of him. I had to reassure him repeatedly that I expected my officers to act like officers, and that I would always provide the resources and—most importantly—the support to thrive in their capacities. I often had to remind him to make his own decisions and stand by them as opposed to being unsure and seeking permission. It was my intent to create a decentralized environment where the officers had the latitude to lead from the front. By maintaining expected outcomes based on formulated

FIG. 2–9. The crew confers at the command post (Photo courtesy of Pat Dooley)

strategies, tactics, training, and policies, my officers had the freedom to grow as leaders instead of being mired in a repressed and confined atmosphere. Understanding the why or the reason behind an order through clear communication gives valuable insight to the person charged with carrying out that order. An effective way to gauge progress is to maintain a vision of what the look and feel of the crew's overall performance should be rather than what credentials and training certifications state it should be (not everything can be learned from books).

Another example of empowerment comes from a senior firefighter who was having trouble with a younger crewmember's consistent lack of initiative and motivation with training, crew cohesion, and overall job performance. The company officer attempted to address these issues with the young firefighter, with minimal results. Eventually, the senior firefighter engaged the young firefighter and explained, in a respectful and detailed manner, exactly what he was doing wrong, what he should be doing differently, how this affected the entire crew's performance, and how, if not corrected, there would surely be punitive actions from the administration. The senior firefighter went on to further explain how poor job performance affected not only the young firefighter's family but his future career success as well. Because of the transformational leadership displayed by the senior firefighter, the young firefighter subsequently improved greatly on all levels. By enlightening, inspiring, motivating, and influencing the young firefighter through the capacity of mentoring, the senior firefighter successfully transformed him into a different mindset and level of performance instead of ignoring him and watching him fail. To that end, somewhere along the line, the senior firefighter was also transformed into the leader he had demonstrated himself to be. Transformational leadership has evolved and is widely recognized as a critical component of the fire service. Firefighters and officers are frequently thrust into unfolding crises of varying degrees and scope, and they need to react instinctively with definitive actions and clarity of purpose without time to ponder and deliberate. By embracing empowerment through the conveyance of authority and the opportunity of others to make decisions, have a voice, and, most importantly, make mistakes and learn from them through self-evaluation opens the mind and unlocks the shackles of restraint.

For some individuals, however, there may be no way to engage in transformational leadership because of their lack of adaptability or a limited capacity for growth. Often, these individuals may remain a participant instead of a leader. You should anticipate this; their fullest potential may never be realized in this arena. A paradigm shift was my end game goal: be a transformational leader rather than the typical task-fixated transactional leader (fig. 2–10). By being path-oriented through providing the proper environment of impetus and

FIG. 2–10. Setting the proper tone with personnel (Photo courtesy of Pat Dooley)

fortitude, I would often remind my personnel, "Don't focus on where you are; focus on where you're going!"

The ancient Chinese philosopher Lao Tzu stated, "Give a man a fish and you feed him for a day. Teach him how to fish and you feed him for a lifetime."[8] This concept is aptly applied in every firehouse, for every crew, on every shift. It's not what they do when you are there; it's what they do and how they perform when you're not there—that's the truest measurement of transformational leadership (fig. 2–11).[9]

Mentoring and Coaching

For the firefighter and the fire officer, education is what you know, but training is what you do! The famous contemporary football coach Bill Parcells is quoted as saying, "Be a player's coach: your best ability is availability!"[10] What does that mean for us? You as the mentor or as an officer need to make yourself

---

8. Lau Tzu, *Inspiring Quotes* (blog), Pioneerthinking.com, June 28, 2013, https://pioneerthinking.com/teach-a-man-to-fish/.

9. Leigh H. Shapiro, "The Dynamics of Change Leadership in the Fire Service," *Firefighting* (blog), Fire Engineering, October 18, 2018, https://www.fireengineering.com/leadership/the-dynamics-of-change-leadership-in-the-fire-service/#gref.

10. Although this quote can be attributed to both Bill Parcells and George Kittle, in its context and how it was stated, I am referring to a quote from National Football League player Todd Gurley in the HBO TV series *The Shop* season 1 episode 3.

FIG. 2–11. Called to assist as mutual aid at this warehouse fire in a neighboring town, our assignment arrives and begins to set up. The tower ladder assigned with us spotted the turntable in between the fire building and the exposure, thus containing the fire to the original building. I asked for a water supply line from a nearby mutual aid engine and was told they were reserving it for a company that had not yet arrived. I articulated to that engine company that the truck company they were reserving it for was still on the highway en route, and that I needed the water *now*! They complied. A simple tactic saved the day. As an incident commander, the focus is on "what do I have, what do I need, where is it going, and what could go wrong?" (Photo courtesy of Pat Dooley)

available to your crew members and others that may be seeking guidance, leadership, mentorship, or simply wish to talk with you. I enjoyed just sitting with the officers and senior firefighters and picking their brains, asking questions, and listening to war stories from years past. The foundation of your position is to teach, train, inspire, influence, and guide your personnel through demonstrated action. In other words, don't just walk the walk and talk the talk: do what you say you're going to do, and be that person you wish had trained and inspired you or who was a positive influence in your career. As with most leadership strategies, buy-in is the key to success. If your personnel don't buy what you're selling, then what's the point? Achieving your goals derives from those you are teaching believing in the messenger, the message, and the purpose. And from that position, it is critical to know and understand your CBA (if you have one) but especially your department directives, standard operating procedures (SOPs), guidelines, or whatever you call them. Without them, it's nearly impossible to hold a position of authority and responsibility and not fully comprehend what is expected of you and your crew members. Within the realm of expectations is the tone you set with your crew or station. Just like the adage about the new convict entering prison for the first time and tuning somebody up immediately to send a clear message that they are someone not to be messed with (mostly in movies, not reality!), so too as the officer, you must set that tone as early as possible. Initiate this by getting all your personnel together and reviewing the department policies, rules, and expectations for conduct. For me throughout my career, utilizing a skillset strategy I refer to as the before, during, and after has been most effective.

The officer usually deals with personnel issues while they are occurring—obviously. However, to be completely effective, it helps to look at issues in this manner: what have you done to set up the environment, so issues have a minimized risk of occurrence; what are you doing about them when they do occur; and how have you set up the future environment, so they don't reoccur? Hence the before, during, and after approach. It is simplistic, measurable, achievable, and really works. I came up with this method because time and time again, for both me and other officers I have worked with, the same question is broached by senior officers and the administration when something goes down: what are you doing about it?! I got tired of hearing this question and came up with this method. This is what I'm doing about it! For example: if one of your personnel did something in the firehouse that was a violation of department policy, you should have already set the tone and expectations within your operational command prior to the violation (a plan), usually by training and reviewing policies (before). When there is an occurrence, your crew is already aware that this behavior is unacceptable and the excuse that they did not know is deemed moot based on them being previously trained. Deal with the situation accordingly when it occurs based on your rules, regulations, and policies, and don't forget to fully document for reference (during). Post incident, review the department's policies and regulations again to reinforce, retrain, and refresh your personnel

FIG. 2–12. This early morning fire in a local bodega is a clear example of how a proficient, disciplined, and focused crew can have an early impact on an incident. My engine company was first due, and the bulk of the fire was knocked down prior to other companies arriving. Getting the jump on a fire prior to other arriving companies was a company theme I employed to demonstrate action, provide buy-in for our mission, and inspire the crew to do their best.

(after). If you fail to address these issues properly, you are simply not doing your job and are actually making things worse (fig. 2–12).

An article by Travis Bradberry talks about "career-killer torpedoes"[11] and I found they aptly apply to our field of study (fig. 2–13). These examples illuminate how firefighters can sabotage their own careers if not careful, and not necessarily in one single occurrence. The aggregation of little things can be just as damaging as one giant screw-up. Thankfully, by being consciously aware of these points, you can successfully navigate through any potential peril.

- **Negativity:** Unfortunately, I can confess that I am guilty of this. The more junk my city and department threw at us, the more disenchanted I became. It's human nature to feel this way. The problem arises when you spread that negativity and it lingers like a foul odor. You can choose how you respond to negativity. Initially its okay to react with emotion; however, as the officer or senior member, you need to be the adult in the room and control yourself with emotional maturity and dignity. Try to get past the emotional component and look directly to the heart of the matter. Only then can you objectively understand something and decide the proper course of action. How you react to something may vastly differ from someone else's reaction, but make no mistake, you are being watched by others, and your reaction will surely

FIG. 2–13. Without self-awareness, career-killer torpedoes can sneak up on you. (Photo courtesy of militaryaerospace.com)

---

11. Travis Bradberry, "9 Things That Will Kill Your Career," *Inc.* (blog), April 28, 2017, https://www.inc.com/travis-bradberry/9-things-that-will-kill-your-career.html.

influence how others react. Composure is a difficult skill that demands a lot of practice.

- **Playing politics:** We've all had the experience growing up of dealing with a teacher's pet or a suck-up. When this behavior manifests into game playing and feeding into the narrative of us versus them (or worse) and decisions are made based on how you want to be perceived or to be in someone's good graces, you are on dangerous ground. Remember, the one person you can always depend on is yourself! People come and people go (chiefs, friends, etc.), but the one constant throughout all of that is you. I was constantly irritating my bosses by reiterating the fact that I was loyal to our mission, not to the man. Yes, I had immense respect for many individuals throughout my career; however, I steadfastly refused to follow them off a cliff and compromise my personal integrity for someone else's bad actions or poor judgement, or for something that someone thought was a good idea at the time.

- **Inflatable ego:** Believe it or not, you're probably not the greatest firefighter the world has ever seen! We are all simply cogs in the wheel of the fire service. Yes, you may have a huge impact and positive influence on many, and there may even be situations and events which propel you to the forefront of notoriety and high regard, but never forget where you came from and from where you started. Maintain control of your ego: accolades are nice, but they have their place. Ask any Harley rider and they'll say, "Chrome won't get you home!" Always have humility, because we are serving a greater good.

- **Fear of change**: The adages, "100 years of tradition unimpeded by progress" and "Firefighters simultaneously hate the way things are and change," ring true. Firefighters both adore and fear change. They fear it because it means being uncomfortable at times and often requires effort and energy, training, and a paradigm shift. This doesn't mean being lazy. It's often simply the disengagement of a skill, tool, or mindset that has been mastered and now may be rendered irrelevant or antiquated. When change is embraced, it happens for basically two reasons. One way is in the context of a new apparatus or similar circumstance which brings about the appeal of newness, much like a kid with a new toy, or a needed improvement. The other way is by force, like when an unpopular SOP is handed down, and on face value it looks like zero thought was put into it, it creates more work, it is more cumbersome, or it fails to achieve a

needed or anticipated goal. Either way, change is a double-edged sword, cutting both ways.
- **Complacency:** "We've always done it this way" is often the muttered phrase when firefighters are presented with a new operating directive. Being complacent isn't just about the fire service, it's more about the individual, or moreover, the tone and attitude of a department's administration. If mediocrity is what you strive for, then you'll have no problem achieving it. This is the 100-pound weight around your neck weighing you and your department down. Nothing ventured, nothing gained, so you do the same old same old just to keep on keeping on! This is boring, defeatist, and dangerously short-sighted. If your department fails to move forward, then you as an individual need to keep moving forward, much like a shark needs to constantly move to stay alive. The old saying, "A pessimist is never disappointed" is aptly applied here!
- **Losing sight of the big picture:** This is not your own personal sandbox to play in and control. You as a firefighter are part of something big, and more importantly, something very critical to modern society. Sometimes you need to take a step back from your own position and understand that decisions are made for the betterment of both the department and the city or town where you are. Budgetary and political decisions are often made for whatever reason, and they may in fact be very unpopular; they may even suck and have a negative impact on you directly, but these decisions are bigger than you are and are made in that context. I always told my unhappy, at times very vocal crews, that if you don't like something, get involved and help to change it. Speak up, get on a labor committee or into a training class, and learn the nuts and bolts of the job so you yourself can have a positive impact other than simply showing up for work. Moaning and groaning about everything is amusing initially but gets old very quickly. What are you doing about it?
- **Low emotional intelligence:** Immaturity abounds in the fire service; it's an inherent pressure release in a job that demands so much of us both physically and mentally. However, when you are the senior member or officer, there are expectations that must be adhered to. The ability to listen to others intently and not flip out immediately when you are challenged or when you simply hear something you don't agree with is another skill that should be steadfast in your toolbox. Maintain your composure, think

through your response, and stay on point! Emotional intelligence is defined as "skill in perceiving, understanding, and managing emotions and feelings."[12] It is indeed a skill; one which needs to be practiced and honed often. Many of us have refined our skillsets and developed a mature mentality and disposition: others have not. This is important for the officer and mentor to successfully develop and tweak because words and actions have consequences, often unintended, and you could find yourself in several unfortunate circumstances, on the wrong side of an issue, or having lost or degraded some level of the hard-fought respect you earned of those around you.

I cite two examples to clarify. The first was when as a company captain, my engine company responded to an EMS call for service in a single-family home. Upon arrival, the mother of the family met us at the door and ushered me and my crew into the living room of the home where we encountered a teenage girl sitting on the couch having what appeared to be a continuous incoherent mental outburst. We immediately focused our attention on applying our EMS skills to her, but the mother informed us that she in fact was not the patient, and this was her normal status due to an existing medical condition. Our patient, the father of the family, was in a back bedroom having some type of diabetic issue, for which we refocused our attention and performed our duties. The problem I created happened when the ambulance crew showed up for transport. When that crew entered the house, they did the same thing we did: they immediately focused their attention to the girl on the couch. Seeing this, I muttered to the ambulance crew "That's not the patient—that's just the sideshow, the patient is in the back bedroom!" Very immature indeed coming from an officer. Unfortunately, unbeknownst to me at the time, the mother of the family was in close enough proximity to hear that comment and she, expectedly, flipped out on us! She berated me and threw our entire crew out of her house forthwith! I had really screwed up! I made our customers mad at us with just a few simple words. We quickly turned the patient over to the ambulance crew and made a hasty exit to the apparatus. As we were walking back to the apparatus, one of my crew members stated he sort of knew this family from himself living in this neighborhood, and he would go back and try to smooth things over. So back in he went while we

---

12. "Emotional Intelligence," Dictionary.com, https://www.dictionary.com/browse/emotional-intelligence.

got on the apparatus to prepare to leave. After about 30 seconds he came running out of the house at full sprint and jumped into the apparatus yelling "Go! Go! Go! Go!" Clearly it didn't go as expected.

Before this EMS call for service was received at my firehouse, I (along with the union president and the assistant fire chief [AC]) was in my office having a meeting about a different personnel issue that needed rectifying. When the call came in, I excused myself and they stayed behind. Upon my return, we quickly finished the meeting and then headed downstairs to the watchroom so I could walk them both out. When we entered the watchroom, the officer's phone line began ringing and I answered it. The voice on the other end said it was for the AC. I handed the phone to him and looked over at the union president and quietly said, "We may have another problem." The AC did the usual phone interaction of "uh-huh, un-huh, mmm-hmmm," then turned and faced me with a half a grin. When the AC hung up, he looked directly at me and the union president and said firmly, "Back upstairs!" I received counseling on the proper way to conduct myself when interacting with the public. A hard lesson to learn about emotional intelligence.

The second example is that of the fire chief and his chief of training at a monthly department computer statistics meeting in which all the division heads were gathered in the Emergency Operations Center to share data, statistics, and other pertinent information. Generally, there are about 40 to 50 staff and line personnel in attendance. When I was asked in front of the group about pump operators and aerial operators receiving proper training for their newly promoted positions, I merely referenced past practice and stated that was generally left to the company officers and members of their respective houses, instead of looking to the training division to teach them how to be drivers, since most often recently promoted drivers have been acting in the capacity of driver for years at their respective fire companies. Well, the training chief jumped out of his chair and started yelling and berating me about my answer, in front of everyone in the room. This went on for a solid 2 minutes before he exhausted himself and sat back down. During all of this, the chief of department just stood there and allowed his training chief to display a level of immaturity and disrespect beyond any form of forceful and impassioned rebuke or rebuttal. This was just plain wrong, and I literally felt insulted and humiliated simultaneously, but just sat there in complete disbelief as to what was transpiring. Then the supporters of the guy who just laid into me in the room applauded when he was done! Unbelievable. To think this behavior was not only acceptable but normal as well was beyond anything I had seen up until now. But to reinforce an already bad situation, the fire chief then stated without missing a beat, "And now we are going to move into our training component of this meeting and talk about emotional intelligence!" I almost fell out of my chair at this point. You as the chief just witnessed this display, did nothing to stop it as a professional administrator, and now you're going to lecture all of us on how to behave? Just one more nail in his coffin as far as I was concerned about his tenure as the department leader and my respect for him as a fire chief and a man! This is not how it's supposed to be. Nobody has the right to act this way toward another employee unfettered and unmitigated.

- **Overpromising and underdelivering:** We've all heard firefighters, officers, or crews boast about how great they are compared with all others. We've all been subjected to tall tales from firefighters and officers alike about how they are somehow the smartest person on the job and everyone else is just plain dumb. We have also been reassured repeatedly at times that some tasks will get done or a certain objective will be met, only to be disappointed repeatedly. As the old saying goes, "When your mouth writes a check that your butt can't cash, you're going to have a problem." When you say you are going to do something, do it! Don't be that person who is so full of it that all you do is talk, all sizzle and no steak, as it's said. It gets to the point that people won't take you seriously anymore because everything you say is not grounded in realistic terms and shaded with a prior history of promising the moon and the stars and consistently coming up short or not at all. To me, bragging displays low emotional intelligence and a high degree of arrogance. Leave the bragging for the kids on the playground (fig. 2–14).

Whether you are a recruit, a new officer, or even a seasoned officer, wherever you are assigned as the new person in the firehouse, you will be tested. Even if you have been transferred in as a reassignment, you will be tested. You will be probed for everything you do. It doesn't matter if you're the best firefighter or officer the department has ever produced, you have a chest full of medals, books have been written about you, and so on! The normal human mentality and response from those who already work there is us versus them, how much can I get away with, and how far can I take it! As the officer, you can take this to the bank—you are being watched, tested, and processed to see

FIG. 2–14. Avoiding or monitoring these career-killer torpedoes can prevent damage to or the ruin of your career. (Photo courtesy of navalpost.com)

if in fact you fit in and if you are achieving the expectations, both good and bad, of the crew members. You should expect this immediately when you walk in. If you are not mentally prepared to engage and overcome, you're going to have a tough go at making this company work for you. I always equated the prison yard story, as mentioned earlier, to this endeavor because it fits: the minute the new prisoner walks out into the general population yard, they should grab the biggest person there and put them down (usually by physical force)! That sends a clear message not to be messed with. It sounds like a great idea, and it works well in movies, but in reality, it takes more skill and finesse. If you choose, for whatever reason, not to take control early in your new assignment, or if you allow multiple opportunities to engage simply slip by and things get comfortable for your crew and firehouse, then eventually your position there, with all your authority and responsibility, will be relegated to nothing more than a handler rather than an officer or leader. Think about that for a minute. You have lost control early, and getting it back other than by force is nearly impossible. You're just handling your crew members, not leading them, kind of like a caretaker, which is not in the officer's job description. Eventually you look around in frustration and ask yourself, "Who's running this place: me or them?!"

As the senior member or the officer, there will always be one person in your crew or station that simply does not fit, whether they just don't get it, refuse to get it, think they are smarter than you and everyone else there, or whatever issues they may have (fig. 2–15). Dealing with this one individual constantly can be one of the most challenging and draining things you do in

FIG. 2–15. As with any crew, there will always be one individual that simply doesn't fit, and after repeated training and mentoring opportunities continues to destabilize and disrupt a cohesive, thriving crew. Sometimes the squeaky wheel doesn't get the oil; it gets replaced!

your entire career. I will talk further in this book about dealing with the problem employee, but for now I want you to recognize that this may be part of your day-to-day dealings. I have always announced at drill time, dinner time, and on occasion at incidents, that sometimes the squeaky wheel doesn't just get the oil; it gets replaced! When all avenues available to you as the officer have been thoroughly exhausted, including mentoring, coaching, and training, there may be the realization either from you or the administration that this individual is simply not a good fit. There is too much disruption, too many discipline issues, and the entire situation is counterproductive. This should never be a case of an individual who rebukes your leadership being terminated because they don't assimilate. It is, however, a last-chance effective tool to maintain crew harmony and effectiveness. This is a demonstrated example of how you can be in charge of your crew and maintain control simultaneously.

The concept and transaction of learning is where you find it, and not necessarily in a preconceived formal environment, and I enjoy sampling those tidbits and using them to my advantage. One such is how some military officers address subordinate members, by the prefix Mr. or Ms. followed by their last name. When addressing officers, it is by rank. This is expected in a paramilitary scalar organization, and I did it when I was required to be formal, such as in public. In the firehouse or alone with the crews, it was by first name, nickname, or given, bestowed-upon-you name! A clear-cut example of this is when the fire chief addresses the mayor as Mr. or Ms. Mayor in front of everyone. Respect, dignity, and the adherence to proper form and decorum is paramount to maintaining a professional department, especially if you are representing that department. I developed a personal style that was both professional and engaging, and it emphasized empathy, grace, and trustworthiness. When I was an officer, at any of my ranks, I would to the best of my ability in public and around others address all firefighters I worked with by their prefix (Mr. or Ms. *insert last name here*) or by their rank (Lieutenant *insert last name here*). There is a time and place for first name and nickname closeness, but in public (and not directly engaged in a situation, scenario, or incident) is not that time and place. Treat others with respect, and you may receive that respect. If you talk like you're in the locker room all the time, don't be surprised when you get what you get! You need to figure out what is the best style for you overall and work on that. If you are a constant chameleon, shape shifter, or weathervane, always changing with the current and popular, that is how you will be perceived and received. Many books have been written about how new officers must change their personality because they are in charge now and the buddy system is not acceptable anymore. This is for you to figure out. Sure, getting promoted to a position of responsibility and authority changes you. But let that change be for the good, not the worse. Embrace this new learning—don't just shy away from it and maintain the status quo of attitude, aptitude, and ability. I will readily admit that not only am I not the smartest guy in the world, but probably not the smartest guy in the room, either. But because I engage my duties in a professional manner, I know I will have to ask for help at times. Asking for help is not demonstrating weakness, no matter what you heard or who told that to you. It's a sign of maturity and humility, and you will get much further in this field of study if you know when to ask for help. I'm not saying every move you make needs someone's assistance. If that's the case, you will quickly become mentally paralyzed, and like I always say, indecision is no decision! For example, back in May of 1989 when I had just over a year on the job and was then eligible to be acting officer of my company (our CBA stipulated after 1 year of probation you can serve as acting officer),[13] I came to work one evening and was told "You're in charge of the wagon tonight!," meaning I was to be acting

---

13. "[Collective Bargaining] Agreement Between the City of Hartford and the Hartford Firefighters Association, July 1, 2009, through June 30, 2016," https://www.hartfordct.gov/files/assets/public/human-resources/hr-documents/hartford-fire-fighter-association-local-760-contract-7.1.2009-6.30.2016.pdf.

officer of my engine company for the shift. Later, we responded first due to a local high-rise for alarms activated which turned out to be nothing. I knew our procedure when responding and what we were supposed to do when we arrived. But when the senior deputy was standing on the sidewalk telling the companies that he was dismissing the assignment, he stated out loud "I'm gonna hold the first-due engine and first-due truck, everyone else is dismissed!" Then companies started to leave the scene and I stood there looking at the deputy, clearly not knowing what I was supposed to do next. As the deputy was walking back to his car, I excused myself and asked if it was alright to ask a question. I thought for sure he was going to light me up and recommend I be fired on the spot (I was very nervous, and he was the big chief). I simply asked him, "Now that my first-due engine company is being held along with the first-due ladder company while every other company was dismissed, what exactly am I supposed to do?! What does that mean?" He smiled and began to explain that if the alarm system was reset and the building is secured, you must wait for everyone else to leave, and then you can go back on the line available (notify the dispatcher that your company can return to service). This chief officer, who played a huge role in how I would conduct myself as an officer eventually, became one of my mentors on the job. He took the time to help me rather than berate or embarrass me, something I will never forget. That's it! Ask for help. Another example is when I was first promoted to officer back in August of 1994. I was a newly minted lieutenant assigned to a busy north-end engine company. I had the requisite KSAs to fulfill my duties, but I found myself unprepared for a lot of what I was dealing with. I would call other officers on the shift to pick their brain, run things by them, ask if this was the proper decision I was making, and just learn as much as I could from them. I knew I didn't know everything, but I knew where to find the answers. This helped immensely, and eventually my phone calls became less frequent because I had learned the necessary information to perform more effectively. All this was accomplished by simply having the humility to ask for help. Don't allow your pride or ego to enable failure. Results, not excuses.

Once assigned to your new crew as an officer, or if you have been transferred into an existing crew but had some experience already being an officer, you will need to do your own evaluation of each member to identify their individual strengths and weaknesses. Who can you depend on to take care of certain situations, scenarios, and functions? Basically, who can you trust to do what needs doing? It also identifies the weaknesses where attention is needed. Training, retraining, review, whatever it takes, but this illumination will reveal areas which require addressing. Likewise, you yourself can be enlightened and recognize the need for extra attention. It's been my professional opinion to not put individuals or groups of individuals into a position in which you know they will fail, especially when lives are at stake. If you continuously ask things of those individuals that you know will fail, you yourself have failed. Work with those individuals to teach, guide, and train them to the standards you hold, and minimize the risk of screwing things up. Things will always happen that you did not expect or anticipate, but if you keep allowing the stinkers in your crew to flop, what does that say about your leadership? Conversely, be mindful that your direction and orders are fair and equal as consistently as possible. For example, if one crew member is considered a solid person and the other in a state of perpetual bungling, do not favor the superstar. This is not fair or equitable and can lead to you being disciplined. How would you explain to your boss the complaint made by one or more of your crew members that you are constantly assigned latrine duty to them, while the other crew members, perceived by others as your favorites, are enjoying an umbrella of immunity from any negative scrutiny or tasks? Testimony like "I'm treated unfairly," or "they don't like me" carries a lot of weight in labor court and could be used against even the most well-intentioned officers. Speaking of an officer making mistakes, it becomes abundantly clear early on that you will need to instill confidence in both your crew members and your bosses. For your crew, they need to see that you know what you are doing; you are confident, competent, dependable, and trustworthy; and you always have their best interests in mind. Your bosses need to see this as well. It's not just impressing those under you. You need to make it your business to instill confidence in your superior officers, so they believe they made the right choice in promoting you. I'm not talking about sucking up to impress them like some high school crush you want to date. I mean showing them you can perform beyond exam time, and although you may need guidance from time to time; who doesn't?! Do your job thoroughly and professionally. Beyond asking for help, as mentioned earlier, one of the fundamental traits of a senior member or officer is knowing where your resources are to gain additional knowledge. I would always tell people I'm not the smartest guy in the fire department, but I know where to find the answers I need when I need them. You as the officer or senior member really need to make a conscious effort to know your resources for any subject that may arise while performing your job (fig 2–16).

When assigning tasks and asking or ordering someone to perform or do something, be careful not to slip into control-freak mode. If you order an officer or a crew to do something, give the order and get out of their way so they can do what you asked. If you constantly stand over them and watch, you are demonstrating several poor behaviors: You don't trust them, you don't think they will perform, and most importantly, you failed to train them properly. Delegating means to instill temporary authority in someone while you still maintain responsibility. If you are going to extend your authority by delegating

FIG. 2–16. Me, conferring with my aide and several officers while at the command post during a working fire. This is command presence in action. Note the paper in my hand: later in this book I will explain the *dance card* method of fireground accountability. (Photo courtesy of Pat Dooley)

something, you'd better make sure they can handle it, because in the end, you are still responsible for that task. If you're not sure, train, train, train! It's not what they do and how they perform when you're there; it's what they do and how they perform when you're not there! Maintaining a coordinated effort to train, guide, and inspire is paramount to successfully achieving your goals as an officer. One thing I have learned from working with all kinds of fire officers throughout my years is that your crew comes first! Place your people above you. That means many of the perks you get for being an officer should be directed at them before you benefit from them. For example, whenever my company received positive recognition for something we did either at an incident or elsewhere, I would always stand in the background and proudly state it was the efforts of the members of the company, not me alone. Teamwork equals pride, and I believe instilling pride comes from strong teamwork. What firefighter doesn't want to take pride in their company and those actions performed as a crew? Another example is something that happened to me directly. I was working overtime on an engine company riding the back step as a private during my early years. It was an extremely cold morning around 5:00 a.m. when we were dispatched first due along with the assignment for a report of an odor of natural gas emanating from a manhole cover down the street from the firehouse. Back in 1989, we still had many old open-step apparatus in the fleet. The officer and driver had an enclosed cab to shield them from the weather's harsh elements, but the two firefighters sitting behind them in the jump seats only had a half roof and no doors, so they were completely exposed to the elements, both good and bad (fig. 2–17). After investigating the source of the odor, my first-due engine company was ordered to stand by and wait for the gas company to respond to mitigate the odor issue; all the other companies

FIG. 2–17. Operating in extreme weather, such as this hot and humid night while operating at a three-alarm fire is at times difficult. Not only are you battling the incident, but you are fighting the weather as well. I frequently pointed out to my crew that we were wearing the equivalent of snowsuits in the middle of summer. (Photo courtesy of HFD)

were subsequently dismissed. We waited for over an hour for the gas company to show up. Sitting in the back of the cab, exposed to the weather temps, wind, and with no heat was rough, at best. My officer at the time was newly promoted and full of himself. He sat in the front seat of the apparatus the entire time, enjoying the heat from the heater and did not even get out of the apparatus to see if me and the other firefighter in the jump seats were okay. We were not dressed for cold weather other than a sweatshirt and extra pair of socks over our regular socks. Back then we had no bunker gear, no hoods, our turnout coats were significantly thinner than they are today, and the gloves we were issued were not made for extreme cold. We were slowly freezing to death while the officer sat in the cab warm and comfortable. I repeatedly asked him how long we would have to remain outside like this, and he basically ignored us both. It was only when I got off the apparatus and walked over to a police car that was blocking the roadway due to the gas leak that my lieutenant asked what I was doing. I told him he either lets me get into the back of the police car to warm up or take me to the hospital for hypothermia! I got in the car. He never got out of the cab to allow us guys sitting in the jump seats to take turns warming up in the cab. Nor did he make any arrangements with the police officer for us to sit in his car, or anything along those lines in terms of looking out for our well-being in this sub-zero weather. Either this fire officer didn't know any better, or he simply didn't care. Either way, this is no way to look out for your crew, or any crew. The officer was selfish, self-absorbed, arrogant, and to me, just plain stupid. I wouldn't allow a dog to suffer that way. The point here is easy to embrace; resentment is poison, so act accordingly when you're the officer. If you don't look out for your crew members, and ultimately screw them over, they will *never* forget that. Firefighters in general never forget when they get screwed over by someone or by something. I say something because if a particular piece of apparatus or equipment has failed to live up to the intended standard, we never forget, and most often stop using that equipment. Same with a lousy officer. That resentment, animosity, and negativity is the only vibe you will get from that officer after they pull something on you.

One more thing about resentment being poison: when interacting with the public, always treat them with dignity and respect, no matter how stupid they can behave at times (fig. 2–18)!

In all my years interacting with the public while in my official capacity, there were times when I simply did not know who I was speaking with, but they turned out to be people important enough to have a direct impact on my career. There have been frustrating, combative, and at times intolerable interactions during which I had to proceed with extreme caution and due diligence to keep my emotions and temper in check. Remember, I'm supposed to be the role model here. Demonstrated action is more positive than just telling someone what to do.

FIG. 2–18. The public! (Photo courtesy of pinterest.com)

Personnel Issues

As the officer of a fire company, why are you here? What is your role or job, and how should you do it? Are you here to blow the airhorn and stomp the Q-siren pedal, then corral the cats like a glorified babysitter for your crew? The answer is in your job description when you applied for the promotion (fig. 2–19). It's in your CBA if you are in a unionized department. It's in the personnel department's rules and regulations stating what each job within your jurisdiction is supposed to do. It's written somewhere, but do you know it?

FIG. 2–19. Sometimes it's your action or inaction will makes things worse when dealing with personnel issues. Be careful and know your job thoroughly, including policy, authority and parameters, and contractual obligations.

When dealing with issues related to your personnel or the people serving in your crew, the best way to stay out of trouble is to handle these issues with the same speed and diligence as you would handle your own issues when they arise. If you constantly dismiss someone's need for assistance or drag your feet because it's not really that important to you, then you are setting yourself up for trouble. As the officer or person in charge, you are looked upon as the go-to member or the problem-solver, not just the leader of the crew or the officer. These are big shoes to fill, but this is the role you are in, the path you have chosen, and the job of the leader and manager. When on the fireground or at an incident, circumstances can be a little different. You most likely have a task to fulfill or orders that were issued that need to be carried out. As an example, I'll use a situation that occurred when I was a company captain. I had just received this new probationary firefighter right out of drill school not a week earlier, and while working one evening we caught a working fire right around the corner from the firehouse. My company was first due, so we pulled out of the firehouse, grabbed the first hydrant we came upon, and stretched our supply line down the street, pulling right up to the fire building, which was ablaze! The probie and I jumped out of the rig, and we both grabbed the 1¾" cross lay and began to stretch down the driveway to the front door of the fire building. About halfway down the driveway, the probie stopped, dropped his bundle of attack line in the driveway, looked up at the second-floor front windows venting heavy fire, and said to me, "You want me to go in there?!" I think he was in shock because nothing at the training academy had prepared him for the real deal. At this point he was frozen with fear, and I had a real dilemma on my hands. My probie was not performing his duties, and I still had to get this line in place and start flowing water on the fire. I picked up the line and stretched it up to the second floor as best I could with him following closely behind and engaged the fire. Luckily there were other companies pulling up to the scene and one of the rescue company guys assisted with my line placement. I completed my orders from the incident commander as expected, because it's about the result, not the excuses. The job needed to be completed regardless of what was transpiring with my probie. My bottom line is always this: if one of your crew members fails to perform, you do it instead (fig. 2–20). You complete the task and worry about the details later. Once things settle down at the incident, you can start inquiring as to what happened and why, and more importantly, what are you going to do about it. For situations that require more than just a conversation, like this one, once you get back to the firehouse, you can begin your investigation as to what happened. Remember to fully document all the information you garner because it may come up later, and the ever-present question of what you are doing about it will inevitably be asked.

The old saying "Nothing kills a good employee faster than you tolerating a bad employee" applies here as well. Many issues and conflicts that arise in the firehouse often stem from the officer's indifference to the legitimacy of the complaint or issue rather than merely a policy failure. In other words, when you don't address issues in real time, people become disgruntled, which leads to negative behavior and distrust.

The Before, During, and After Model and Handling the Problem Employee

By utilizing my tried-and-true method of dealing with personnel issues, I utilize the following steps of the before, during, and after (BDA) model that I

FIG. 2–20. Me, operating this handline assisted by the rescue company member. It's about the results, not the excuses. Do the job, complete the task. (Photo courtesy of Pat Dooley)

employed for both fireground incidents and incidents in the firehouse, and wherever else they may occur:

- **Before the incident:** What policies, planning, and training elements are already in place that directly address this incident, situation, or scenario if and when it occurs?
- **During the incident:** First, complete the task! Whatever is happening, the task still demands completion to fulfill the assignment requirements. Afterward, when things calm down and you have both the time and the appropriate setting, initiate your investigation. Figure out what happened and more importantly, why it happened. Develop a plan of corrective action and implement it. It should be closely aligned with department policy, but more importantly, it should directly address the needs of the situation and scenario, specifically your needs. What do I need to do to fix this situation?!
- **After the incident:** Document all your interaction thoroughly for future reference and to demonstrate actions taken by you to correct this behavior, as well as trends, repeat occurrences, and similar situations. Department policy must now be reiterated to the entire crew as a sort of refresher and reminder that certain behaviors are expected of them.

At this point, you have addressed this type of behavior that occurred at the incident before it happens, while it's happening, and after it's happened, hence the BDA model I developed from experience. Developing a plan or corrective action can be as simple as telling someone to pay closer attention to details or to knock off a certain behavior. My dad used to tell me to "cut the crap!" This can get very complicated, so after falling flat on my face repeatedly, I've developed a process to help achieve my goal:

1. Ask the individual what happened and to explain their logic as to why they behaved a certain way or did something unacceptable.
2. Dissect their response: Did they take ownership of their actions? Are they unwilling, unknowing, or unable to perform to expectations, and why? What is the root cause of this behavior? Do they require engagement with the Employee Assistance Program (if your department does not have a program, most likely your city or town jurisdiction does)? Some people overlook the EAP option because an individual in personal crisis and presenting a masked cry for help may be perceived as displaying

negative behavior to those not paying close enough attention to their personnel.
3. Are there clearly defined expectations set by you or the department which serve as guidance and guardrails for behavior and set expectations?
4. Role reversal: Perform a role reversal exercise with the individual by asking them what they would do if they were in charge, and you did what they did. Often, when seen from a different perspective, behavior is easier to understand and to apply positive change, especially when the authority and responsibility is shifted to that individual. We have often heard statements like "If I was the boss I would…"; well, here's your chance to demonstrate!

Formulate a corrective action plan, whatever that may be; however, it must be simple, measurable, and achievable. Too many moving parts, tasks, or people involved serve only to complicate things and can end up slowing or shutting things down, often out of frustration. Can you measure your results? Did you go from A to B with your plan of action, and can you demonstrate sustained results? Is your corrective action plan achievable, or is it just based on theory, ideal but unrealistic conditions, or even an alternate reality? Does your plan match the learning style of the individual you are working with? There are many variables to effective communication, and you need to be aware of these and possess the ability to recognize and adapt your actions and plans when required. In my department, we have many personnel whose backgrounds are from all over the world, so when I speak to them, can they understand the language I'm using? I often speak with jargon, shortened words and phrases, and rough unsophisticated verbiage, and some people may not understand what I am trying to communicate. Is the cadence of my speaking too fast, and they are missing key words because they are simply trying to process all they hear? Is there a cultural difference between me and the individual which may interfere with understanding, or perhaps an educational deficiency which precludes them from understanding certain concepts? Is their overall background a factor in creating a comprehensive communication barrier? Just because they were hired onto the department and passed through the fire academy does not necessarily mean they are all geniuses; it takes work to be an effective communicator, especially as an effective officer! Best advice: know your audience! Explain in detail the significance of why it is so important to follow your direction as the officer as well as department policy, especially in this context. By revealing the why, you allow the individual to thoroughly understand the importance and gravity of orders and direction issued to them. No mysteries, just thorough, comprehensive communication. Review the disciplinary process with them and explain how they fit into the progressive discipline process at this juncture. Progressive discipline is often defined as a verbal warning, a written warning, suspension, then termination. At this juncture, we are simply revealing the structure of progressive discipline and where the individual falls within. Document this interaction as a counseling session. If future incidents occur, then we escalate up the progressive discipline ladder, obviously with the fire administration taking the lead. Reassure the individual that you fully support their best interest and more importantly, their success. Without this, animosity and resentment seep in and start to grow. You need to fully commit to your personnel. Follow up by keeping a watchful eye on the individual's behaviors in the immediate future to ensure compliance and proper conduct. Properly handling the problem employee can be the challenge of a career for a fire officer and often produces the most frustration (fig. 2–21).

President Franklin Delano Roosevelt once said, "Motivate [people] by appealing to their aspirations and their fears!"[14] I have developed a strategy to do just that. There are two elemental sides to this strategy: the professional side and the personal side. First, the professional side includes the progressive disciplinary process consisting of an oral reprimand, then escalates up the chain to a written warning, a suspension, and then termination (obviously the fire administration would be handling this part) if no corrective action is adhered to. This is basic officer procedure when dealing with personnel. This can be a game changer for someone who is a difficult individual within their respective agency or department. Punitive discipline can negatively impact your paycheck, your aspiration for future promotions, and your record, as well as the time you serve on the job.

FIG. 2–21. The incident commander conferring with the chief's aide regarding an on-scene personnel issue. Often it can be handled right then and there, but sometimes issues require escalation and documentation. (Photo courtesy of Pat Dooley)

---

14. Quote from Franklin Delano Roosevelt's character in Showtime TV series *The First Lady*, season 1 episode 9, "Rift."

Within my department there are stories about firefighters who were seeking to retire but had to make up their bad time first. In other words, to qualify for their specific pension, they needed a certain number of years worked. Being on suspension days does not equate to worked time, so those days someone served while on suspension needed to be made up to fulfill the pension qualifications.[15] Many individuals did not realize that suspension time off early in their career needed to be made up at the end of their career, thus throwing their retirement plans into a tailspin. All avoidable with simple good behavior.

Second is the personal side, which can weigh equally heavy on an individual who falsely believes there is nothing to fear. There is a huge family impact on someone who is on suspension, or worse, terminated from the job. Loss of pay and income is obvious, but how do you explain to your family members that you don't have to go to work this week, and it's not vacation time? Firefighters are inherently viewed and often aspire to be seen, especially by family members and children, as a role model. How then do you explain your punishment to them? Additionally, elders of the family may see you as a role model as well; do you want to embarrass and possibly bring shame on them? You may end up with a negative record to follow you around for the remainder of your career, and more importantly, a negative reputation which ultimately effects your image and identity as a firefighter, which can equally impact your aspirations and arouse your fears. I frequently say that your reputation is a commodity which needs to be protected, cultivated, and utilized wisely and with humility (fig. 2–22).

FIG. 2–22. Me at a working fire having to explain to the incident commander why my assigned task did not go as planned. Although I stretched the attack line to the front door, I expected the fire to be further inside the structure, so I didn't flake the line out. Turns out the fire was in the foyer by the front door, so my handline turned into a pile of spaghetti blocking the building's entrance. (Photo courtesy of Pat Dooley)

---

15. "[Collective Bargaining] Agreement Between the City of Hartford and the Hartford Firefighters Association, July 1, 2009, through June 30, 2016."

Sexual Harassment

Let's dive into a topic that many believe is radioactive and fear dealing with because it can go wrong in so many ways. These intrepid feelings are reinforced from hearing and reading about horror stories from other firefighters and incidents. Nevertheless, as the senior member or the officer, you must deal with the dreaded sexual harassment complaint. For this writing, I am going to use a female firefighter and a male firefighter for context, but sexual harassment, and harassment in general, knows no boundaries and is unacceptable at any level and in any form. In today's modern fire service, the things said and done 30-plus years ago are no longer acceptable today. Our society, people, and circumstances are much different now, and sexual harassment finds no home in the fire service, yet believe it or not, it still occurs, so you not only need to know how to handle it, but also how to effectively mitigate a potential disaster.

Featured in a 1983 recruitment brochure[16] that was widely disseminated at school and media events to generate female interest in the fire service, these two pioneering women, Maria Ortiz and Zandra Clay, were the first female firefighters hired by the City of Hartford in 1982, and the first throughout New England at the time (fig. 2–23).[17] The department was woefully unprepared for their entry into the ranks of the fire service and many challenges and changes lay ahead for this progressive department. They faced many struggles, first and foremost being the attitude and disposition of an all-male department in the early 1980s. The department didn't really modify the bunkroom and bathroom facilities, and the firehouses remained unchanged for many years. The female firefighters were asked to wear sweatpants and sweatshirts while in their bunks. These women just had to deal with fitting in. Couple that with an inherent animosity in comments such as "this is all-male job," "women don't belong here," and "women can't do the job, and aren't strong or tough enough," and you have a recipe for trouble. Even many of the wives of firefighters complained loudly about their mere presence in the firehouse, especially in the living quarters of their assigned stations. But the administration and the department persevered in this endeavor, and eventually more and more women entered the HFD ranks of service.

For context, I will speak of harassment that is considered unwelcome behavior, and although there are many forms, definitions, and legal terms that apply, the one referenced here is what happens with frequency in the firehouse. If you are the officer, the senior member, or simply an acting officer for the shift, you really need to know how to handle these situations with as much confidence

---

16. Hartford Fire Department, Recruitment Brochure, 1983.

17. Hartford Fire Department, First Female Firefighters, 1982.

FIG. 2–23. Maria Ortiz and Zandra Clay pictured in a recruitment brochure from 1983

as you would a fire or rescue situation; this is equally important and bears the same gravity. As the interim AC, I attended a daylong conference at a prominent law office in downtown Hartford. This law firm specialized in personnel issues, and they even had a mockup of an actual courtroom in their conference center to train lawyers on how to behave in a courtroom. Along with my boss (the chief of department), there were many other career chiefs from various local departments in attendance as well, and the conference was geared toward guiding department administrators through various legal issues which may arise in any department. The emphasis was on how to handle these situations to mitigate any negative outcomes, and to successfully navigate through the various issues that present themselves when dealing with this stuff. One term they emphasized was the concept of *first breath*.[18] According to the law firm

---

18. Shipman & Goodwin LLP, "First Breath: Sexual Harassment Reporting," Professional Conference of Connecticut Fire Chiefs in Hartford, CT, 2013.

which deals with all kinds of legal personnel issues, first breath is the term to indicate that if someone in authority is a mandated reporter or in a position of responsibility and they become aware of a sexual harassment situation, whether they were looking for it or not, if they hear of it, they own it! In other words, as soon as you hear the first breath of this situation, you must act immediately. The premise for the law firm instructing the chiefs on how to proceed was simple: consider everything you and your department does to be seen in the context of being in a courtroom and having to defend not only yourself, but your department as well, so handling it properly the first time is paramount to success.

I like to play this episode out as if it were to occur in the firehouse, as if I was the officer, because in fact, this has happened to me several times, and each time I learned something new to improve my response and mitigate the situation the right way. Now I'm going to pass this on to you. This is not *the* way, only *a* way, so you do what you think is best. Here's the scenario: firefighter (FF) Smith, a female, comes to you the officer and states that FF Jones, a male, has done something, said something, or has offensive material visible from his locker or phone, or whatever the case may be. She says it's not that big a deal, she only wants him to stop! She is disinterested in getting him or anybody else in trouble because, for obvious reasons, she must work with these guys and doesn't want to be labeled or gain a negative reputation, whatever that may be. You see the dilemma I'm beginning to lay out here? Contrary to the most blatant of reported complaints, this *soft complaint* is the most frequent, the most engaged type, and the most dangerous in terms of not being handled properly. Just because it's a soft complaint doesn't mean it's not a complaint. If someone says they're having crushing chest pains, but it comes and goes, would you dismiss that complaint as a soft complaint? I'm going to reference the EMS comparison more frequently here to provide my usual context and perspective to help guide you in understanding what exactly is happening here. Also, beware that just because a complaint is not made, if you see something funky, you should act on it. Time and time again, I have heard horror stories within my department about an officer not doing anything or not doing enough to mitigate issues, and that officer being on the hook for liability, discipline, and consequences. FF Smith informing you she just wants FF Jones to stop (whatever it is he is doing) but doesn't want to necessarily get in him in trouble is her corrective action. She should be the one telling FF Jones, "Cease and desist, otherwise I will escalate." If this is unsuccessful, then she can come to the officer for further assistance; however, that would be her first recourse to remedy this situation. But that doesn't always happen, does it? The first thing you should be doing when a situation like this comes to your attention and grasps your focus is the same thing an emergency medical technician (EMT) would do while at the scene of a trauma call: stop the bleed! You as the officer need to stop this negative action immediately by taking mitigating measures now. Not later, not "I'll investigate," not "I'll talk to him" or any other ambiguous inertia. *Now!* Whether it's to separate the parties involved or to take the company off the line temporarily, you must intervene and stop this action immediately. Think of the implications if you do not. Then notify your chain of command immediately. For a company officer, notify your shift commander. If you are a shift commander, notify your administration chiefs immediately. Do not wait. The gravity and enormity of this situation cannot be understated right now. If you think I'm making a big deal of this unnecessarily, how would you respond under oath in court to being questioned as to what exactly you did (or didn't do) as the first reporting officer or person in charge? Next, engage your department policy, whatever that may be. Pull it out of the notebook or computer, sit everyone down in the kitchen, and review the policy, what is expected, and what the exact parameters of responsibility and authority are for each firefighter. Then, as expected, document this event thoroughly. Who, what, where, when, why, write it all down, not only for your official record, but for future use, because you will surely be questioned about this in the future. Your fire administration needs to handle this situation beyond what you have already done, which is to stop the bleeding and notify your chain of command. Keep in mind, however, that if you do nothing, your inaction will no doubt lead to legal action against you, your department, and your city or town. Besides, what if a female firefighter came to you and told you of a situation that needed rectifying, and you did nothing? What do you think she would be doing immediately after you placated her and didn't take her seriously, or worse, didn't take any action? She would most likely be on the phone in a quiet corner of the firehouse or outside in the parking lot, either with her best friend or her mother, telling them your failure to act, saying, "He didn't do $#*% for me!" Handling this type of incident is relatively simple and forthright. We unfortunately often make it difficult by not handling it properly. You obviously need to support the impacted individual and render all necessary care to insure their physical and environmental safety. If need be, take immediate steps to mitigate the situation and defuse a possible escalating incident. However, your official investigation should be limited to what transpired and what you did about it in the immediate. Any further probing should be conducted by the administration.

Assume several years down the road you are hauled into court and are a witness or defendant under oath on the witness stand being questioned by the plaintiff's attorney in this matter. When asked what exactly you did to handle this situation, and you begin to muddle and mumble your way through what is supposed to be a cogent, definitive response, you surely will be asked a few follow-up questions including the following:

> *What legal authority do you have as an officer to investigate a complaint of sexual harassment?*
>
> *What specific training have you received, including certifications and refresher training, to be qualified to handle this type of incident?*
>
> *What exactly does your department or city policy state your actions are when handling a complaint of this nature?*
>
> *What training either in the fire academy or anywhere else do you possess that qualifies you to investigate and mitigate this type of complaint?*
>
> *How many incidents of this nature have you handled, and what were each of their legal dispositions?*
>
> *Have you ever been disciplined for failure to perform your official duties as an officer or mandated reporter?*

I would most likely fold under that much scrutiny in court. I'm simply not qualified to handle this type of situation properly beyond what I do to stop the actions and render the individuals safe, which is expected of me by policy, and then immediately kicking it upstairs to the administration. They are trained, they have the resources, and they have the authority to deal with these types of incidents. Although this may sound like upward delegation, it is simply a case of maintaining official parameters of the scope of your authority. If you were an EMT on a trauma call, and you successfully stopped the bleeding patient from complete exsanguination, would you then whip out your trusty tool kit and perform street surgery on the patient? Of course not! You're not qualified, trained, or skilled to do such things, and it's not expected of you to perform those tasks. So why do you think it's acceptable to dissect a sexual harassment complaint and handle it to completion? Again, you're not trained, qualified, or expected in your official capacity to do these things; other than stopping the bleed, notify your chain of command and let them handle it! Afterwards, as the company officer, you should engage my tried-and-true way of handling issues with the BDA model referenced previously. Before an incident like this occurs, everyone should be trained and fully aware of the department policies and what actions they can expect to be taken. When the situation does occur, what should you be doing to mitigate the situation? And after the situation has been mitigated, what training do you provide to reset the environment and reaffirm prevention so that another similar situation does not reoccur? I'm not telling you this is foolproof; all I'm saying is after I have made multiple mistakes in my career, this is the most effective method for handling these types of incidents. Remember, indecision equals no decision, so beware!

Case Study

My very first shift out of drill school (what the department's fire academy was referred to) was a middle shift of a three-night trick. Back in 1988 we worked a three-on, three-off work shift, rotating days and nights. After my initial first two weeks in the firehouse, the rules stated that probationary firefighters were to be placed into rotation for detail, meaning you work your regular shift, but you would be temporarily assigned for one, two, or all three of your working shifts to a different company in a different firehouse, wherever the vacancies were for that work shift. Even though we had a minimum manning compliment of 80-plus members per shift back in those days, not everyone that was scheduled showed up to work. Some were sick, injured, or on accrued time off and other excused absences, so if an apparatus could only ride four personnel but six were assigned and showed up to work, two would be on the road, detailed to another house where the vacancies were. When I walked into the firehouse to report for my night shift, the lieutenant said I was detailed down the street to a single engine firehouse. I gathered my gear and my bedding, stuffed it into the trunk of my car, and off I went. When I got there, I reported to the officer, and he told me to relieve the on-duty hydrant man and that was my riding position for the night. I spoke with the hydrant man, secured my gear, and he left. Now I was looking around asking myself, "What do I do now? Is there something I should be doing at this firehouse? What does it mean to be detailed?" From the watchroom, the senior member approached me, introduced himself, and began to show me around, point out where everything was, and tell me what to do when this or that happens. He was a lifesaver as far as I was concerned. I quickly settled in for the night and started my expected duties.

Questions

1. Although the officer is present, what role does the senior member play, especially with probationary firefighters?
2. How can the senior member and the company officer accomplish their respective duties effectively without interfering with each other?
3. What is the difference between the officer and the senior member, other than rank?

3

Succession Planning: Promotional Preparation and Advancement

This chapter, although aptly named "Succession Planning," is based in the framework of what that means to the line or staff firefighter or officer, not the fire administration. The more common application of this term refers to the leadership of the department preparing subordinates to replace them in succession when the opportunity arises; however, the term is not exclusive to a fire administration. To me, the term *succession planning* is just a fancy way of saying I want to get promoted! But as Antoine de Saint-Exupery stated, "A goal without a plan is just a wish!,"[1] and he would be correct in that statement. Years ago, departments preparing candidates for promotional opportunities used to call this process *grooming*. Today, that term carries a negative connotation (use your imagination) and is widely out of use, having been replaced with the new and improved term *succession planning*. The actual definition of succession plan is to ensure the continued effective performance of the department through strategic recruitment, development, and replacement of key personnel.[2] For us, just exactly how would we employ this plan to successfully gain that promotion we have our eye on? There is more to it than just filling out an application. The first thing you need to know is what your goal is. What aspirations do you have, what career path do you want to proceed down, and what interests you and why? Some people are merely interested in the pay bump, while others seek out more responsibility and authority so they can have greater input and a stronger impact on the department. As with the scientific method,[3] you then need to define the need. Does your department frequently promote, or almost never?

---

1. Antoine de Saint-Exupéry, "A Goal Without a Plan Is Just a Wish," ed. Meet Shah, SetQuotes, January 22, 2022, https://www.setquotes.com/a-goal-without-a-plan-is-just-a-wish/.

2. "Succession," Dictionary.com, https://www.dictionary.com/browse/succession.

3. *National Fire Protection Association (NFPA) 921: Guide for Fire and Explosion Investigations* (Quincy, MA: NFPA, 2021).

Are there existing or anticipated openings in the job that you are seeking, such as driver or officer? You will need to do some research within your agency to determine what opportunities are available or will be available. For example, I was only interested in seeking promotional opportunities for officer positions; however, I tested for every exam that I was eligible for, such as the driver jobs, fire marshal jobs, public educator jobs, and whatever else I could get my hands on that my department was posting a test for. The reason? I was building toward my goal to get the promotion I wanted (fig. 3–1). Every test I sat for was a practice exercise for me. All the associated studying I did was building my body of knowledge and study skills so eventually when I took more tests, most of the material was rote. Taking all those exams made me a better test taker, and I was able to see the same questions repeatedly on multiple exams. Another equally important element is mentoring. Seeking out the proper mentor to assist you in getting your promotion is important, but equally important is for you to be a mentor as well. The more you can review the materials being tested, in both directions—to you and from you—the better off you will be at test time, and you and someone you mentor will have gained knowledge that

FIG. 3–1. One of the proudest moments for myself and my family: my mother pinning my gold chief's badge on my uniform. My only regret is that my father was not alive to see this happen. Pictured with me is the chief and his two assistants, along with the mayor. This was the pinnacle of my career, with all the hard work and tireless efforts paying off. Now, all that was left for me was to live up to expectations and ensure that I performed my new job correctly and served with humility and respect. (Photo courtesy of Anthony Williams)

otherwise would not have been readily available. This is the win-win scenario, so to speak. Who to ask for mentoring and where to find the materials needed for the exam is critical. When I first got on the job, the department did not post reading lists, just the exam announcement. Years later, they posted a reading list with all the books the exam was extracted from, which obviously made life a lot easier. But don't just leave it to the books. Know where to find all resources necessary to fully comprehend the concepts and theories of what will be on the exam. Leave nothing to chance. If you believe there will be questions about personnel on your officer's exam, it would be a good idea to go over to your human resources department and start asking questions and seeking materials that can help you. Your reading list may provide a set of specific books, and then state something like department policy A, or rule and regulation B, or something that is not readily available to you. You need to find it.

Another component is your educational element. When I first started in the department, college was looked upon as an unnecessary element. Today, its critical to higher promotion and effective leadership. Whatever you decide regarding seeking a college degree, make sure that the institution you graduate from is regionally accredited from one of the five nationally accrediting regions.[4] If your degree is not regionally accredited, it's basically worthless. I found this out the hard way, so before you enroll and plunk down your retirement fund to pay for it, make sure the school is regionally accredited. Go online to both the school website and the U.S. Department of Education, and you'll find what you're looking for. To develop the path to your future, as I refer to it, you will require certain tools, one of which is a resume. This will be of service to you, and not necessary for your department. Most departments, when they post a promotional announcement, require you to submit materials to prove you have training and education. Although some departments require a submission of a resume for higher ranking positions, others rely on the actual job application to indicate what training, education, and experience you may have, and ask you to provide proof at a future date during the promotional process. For the private sector, handing in a resume is core to the hiring process, but varies in the fire service. The fire service is technical based in its training and education and focuses more on what you have done as opposed to what your document says you have done. However, for you, the resume has another function: it serves as a summary and timeline of your accomplishments, not merely a list of where you worked or what classes your butt was in a chair for, but actual when and where achievements. If you don't already have a resume, you can create one

---

4. Database of Accredited Postsecondary Institutions and Programs, U.S. Department of Education, https://ope.ed.gov/dapip/#/home.

and edit as often as necessary. Your resume is measurable because it is a living, breathing entity, and I often edit mine to be current and relevant (fig. 3–2).

Its purpose is to serve you for future reference, so you can go back and see all your accomplishments in one place, rather than try to figure out what you've done and where you did it. It is an actual tracking roadmap of your professional career (fig. 3–3).

LEIGH H. SHAPIRO

18 Nursery Drive | West Hartford, CT 06117 | 860-523-5077 | bernenbush@comcast.net
LinkedIn / Facebook / Instagram / YouTube / TikTok - LeighHShapiro

CHIEF OF FIRE DEPARTMENT

As a 40-year veteran of the fire service, I am an accomplished public safety leader with extensive experience in emergency operations, planning, goal setting, program development, and negotiations. My well-earned reputation as a bridge-builder and problem solver providing leadership, management, and organizational development distinctly qualifies me to lead a department. During my career, I have established a strong progressive leadership culture, initiated the first officer development program, expanded personnel diversity and recruitment, and elevated training and capabilities for the department's heavy rescue unit. As Second in Command of a progressive professional modern department, I pursued benchmark standards of coverage and technology inclusive to all stakeholders to effectively and efficiently mitigate fire protection, emergency management and emergency medical vulnerabilities and risks. Through a forward critical-thinking vision, I developed and implemented strategic plans incorporating innovative strategies for fire prevention, fire suppression and rescue services, emergency medical response, risk management, disaster preparedness, and emergency management and recovery, providing superior fire protection and emergency medical services delivery to meet the current and future needs of the community.

Areas of Expertise

- ✓ Leadership
- ✓ Change Management
- ✓ Crisis Management
- ✓ Emergency Management
- ✓ Operations Management
- ✓ Problem Solving
- ✓ Public Policy
- ✓ Incident Command & Control
- ✓ Conflict Resolution
- ✓ Human Resources
- ✓ Strategic Planning
- ✓ Emergency Medical Services
- ✓ Communication
- ✓ Disaster Response
- ✓ Municipal Budgeting
- ✓ Occupational Health
- ✓ Project Management
- ✓ Training
- ✓ Tactics
- ✓ Public Speaking
- ✓ Preparedness

PROFESSIONAL EXPERIENCE

ADJUNCT PROFESSOR / INSTRUCTOR 2018 - PRESENT
- Adjunct Professor- University of New Haven - Fire Science and Emergency Management Degree Program, West Haven, CT.
- Adjunct Faculty Instructor - Gateway Community College - Fire Technology and Administration Degree Program, New Haven, CT.
- Adjunct Instructor - State of Connecticut Office of Education and Data Management (OEDM) - Fire Investigator Pre-Certification Program
- Fire Service Advisor, Professional Outcomes - Capital City Industries, Hartford, CT
- Instructor - Fraternal Order of Leatherheads Society (FOOLS), New England Chapter

FIRE SERVICE CONSULTANT   WEST HARTFORD, CT 2016 - PRESENT
- Consult with municipal and volunteer fire administrations, as well as provide individual mentoring as subject matter expert for strategic and technical matters. Assess infrastructure, risk and safety needs and provide administrative, financial and policy expertise. Provide expertise in oral/assessment exam readiness and succession development through individual and group lectures.
- FDIC International 2021,22,23,24 Presenter of Training Course titled Critical Thinking for Fire Officer Development: Readiness Skillsets for Effective Mindset, Approach, and Delivery.
- Principal / Special Expert - NFPA 1710 Fire and Emergency Service Organization and Deployment - Career (FAC-AAA) committee.
- Senior Technical & Strategic Advisor - Capital City Industries, Hartford CT.
- Reviewer and Contributor for Jones and Bartlett fire service textbooks.
- Contributor to Fire Engineering Magazine. [Published]

FIG. 3–2. This is an example of a cover sheet for a professional resume. It is formatted to create a coherent, cogent understanding of what the individual has accomplished. There are many services available to have yours written properly. Although most resumes are one or two pages, the science-based fire service demands comprehensive details of your record illuminating training, education, and experience, and may be several pages in length.

FIG. 3–3. This long and winding path is a metaphoric depiction of your resume and your future direction. (Photo courtesy of Johannes Plenio on unsplash.com)

I had my resume professionally written years ago, and all I do now is plug in the new material as needed. Although resume writers and private sector employers will tell you that a resume should only be one or two pages long, remember, us firefighters must build our professional status with training, education, and experience credentials, so as they say, if it's not written down, it didn't happen! We are in a profession dictated by science, so having thorough and comprehensive details of your fire service involvement is vital. I literally put everything I do on my resume to demonstrate a consistent pattern of professionalism and dedication to the fire service. Now I can see exactly what I have accomplished and achieved. When I was in training for my Connecticut fire marshal certification, one of the expectations for us was to build a curriculum vitae (CV). The Latin term *curriculum vitae*, meaning the course of your life, commonly referred to as a CV, is the extensive and all-inclusive summary of an individual's education, training and qualifications, and publications.[5] Its purpose for a fire marshal is to serve as proof that in fact you are a qualified witness when summoned to court to testify under oath regarding a fire investigation you conducted or were part of. It generally consists of a notebook with all your information inside, like a three-ring binder (fig. 3–4).

I had never even thought of creating a CV for myself up until fire marshal school, but the concept was very intriguing to me because it serves a practical and useful purpose. It is your body of work you have generated throughout your professional career. It clearly demonstrates all your achievements by showing you, not just telling you like a resume, and is all inclusive. It has everything you've done. The best part: it's portable! You can bring this binder wherever you need to prove you have certain qualifications and credentials. Instead of having framed wall hangers at home covering up the stains and holes in the wall, I gathered all my certifications and put them in a binder. Now I know for sure where everything is, and I can see what I have accomplished.

---

5. "Curriculum Vitae," Dictionary.com, https://www.dictionary.com/browse/curriculum-vitae.

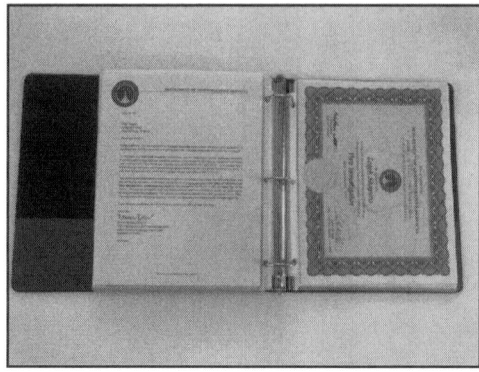

FIG. 3–4. This CV is set up with the score sheet on one side with the accompanying certification on the other and broken down into categories: emergency medical services training and certifications, fire training and certifications, and fire marshal training and certifications, along with any published materials which illuminate professional competency.

Unlike a resume which basically tells someone what you have accomplished, a CV is the actual work that you can show someone, thus proving its legitimacy. For me, the resume is the path to your future, but the CV is like all the keys on a janitor's keyring. Each credential in our CV is a key on that ring, and each one unlocks a door for you, creating potential opportunities for your future (fig. 3–5).

While we are talking about getting promoted, I want to provide some immediate takeaways for you to minimize your stress level during this process, more specifically, during the oral exam. As anybody who has taken a promotional exam before can attest to, the written portion is basically one big memory test. How much information can you jam into your brain and then release for the exam? For my department, our standard testing process consists of a written exam and then an oral board panel. All my oral boards were in front of a three-person panel consisting of various ranking officers from different fire departments, and they would actively score you during your responses. Your department may conduct assessment centers as the exam, but within those assessments are several oral-board-style interactions that you must successfully complete, so this information is still relevant whether you're facing an oral board or an assessment center.

First, once you are notified that you passed the written and will face the oral, or if you are notified that you are facing an assessment center, you must know the job for which you are testing, period. Where do I get this information? From the job description in the exam posting. From your labor contract; there should be a description of each job and their duties and responsibilities. From your city or town hall human resources department; they have all this

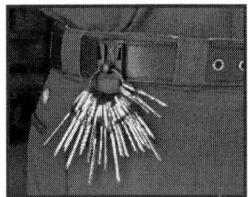

FIG. 3–5. Each document in your CV is like a key on a janitor's keyring. (Photo courtesy of key-bak.com)

stuff documented, it's what they do. Check your rules and regulations book for information. Wherever you find the information, remember, this is what you are being tested on. If you don't know the exact parameters, duties, and responsibilities of the job you are testing for, how will you know how to answer the questions? Know the job you are testing for inside and out.

Second, deliver your answers as if you already have the promotion, have been doing the job for some time, and are simply replying to the questioners as if they were visiting fire officers from another department and were inquiring as to how you would handle this situation or scenario. This immediately puts your mind at ease because now you think you are talking to other firefighters, not an oral panel. When I was testing for deputy chief back in 2010, I came across an article in *Fire Engineering* by Chief Anthony Kastros from the Sacramento California Fire Department. One of his recommendations for oral exams was to get your head into this mindset that you already have the job, and you're just telling others how you do it.[6] This is probably the most profound advice I ever received regarding an oral exam. It immediately put my mind at ease for this process and changed my trajectory, and by employing this and other techniques, I ultimately scored number one on the deputy chief's test and got the job. I called him after I got promoted to personally thank him for that article, and years later when I attended his class at Fire Engineering's FDIC (Fire Department Instructors Conference), I thanked him again. I believe this is paramount to your success, and that's why I am passing it along to you.

The third takeaway I offer requires more effort and discipline, but once you start to do this, you will see how valuable and important it is for your exam success. Answer every question you are asked like you're in the actual oral exam. I don't mean to act like you're in an exam, with posturing and modulating your voice. What I'm referring to is the answer you give when asked a possible or potential oral board question. You will recognize certain questions more and more, especially if you start taking every test that's offered to you, like I recommended earlier. The more you test, the more you will see certain questions, and the more you will become comfortable answering these

---

6. Anthony Kastros, "Acing the Oral Exam," *Fire Engineering*.

questions. But let's take it a step further, and view someone asking you a question like it's free practice for your upcoming exam. Study groups and sessions are good, but not always available or convenient. But answering a question like you would at the exam is plentiful, frequent, and gets your head into the proper mindset. The first example would be if you were conducting a training session, or if someone you are mentoring asked you a question about how to do something. Instead of simply blurting out this or that like you normally would, give a thorough, well-rounded answer with details and specifics, just like you are being tested at the oral. On a fire marshal oral board exam, the question is often asked "What is fire?" If you simply answer something like, "It's the red stuff we put water on," you will probably get a low score. However, if you answer with, "Fire is the rapid oxidation of fuel in the exothermic chemical process of combustion releasing various degrees of heat, light, and byproducts," you will score high, which is your goal. I know it's a mouthful to remember, but if this question or something similar pops up, give this same answer each time. First, it provides free practice for your delivery, and second, it provides muscle memory for your brain, until the only way you know how to answer that question is with that exact, correct, oral-board-worthy response. Another example is when I was prepping for my chief's oral exam. During that time, I was an engine company captain, and one summer afternoon we responded to a food-on-the-stove type of call in an adjoining neighborhood to my district. Our usual assignment of three engines, two ladder companies, and a rescue and a chief officer responded and ultimately mitigated the incident. While we were picking up our tools and repacking the hose onto the engine companies, an older woman walked up to the group of us firefighters and started to ask questions, in an almost pointed manner, as if she believed we were doing something wrong. As the ranking officer in the group, I stopped packing the hose and politely engaged her in conversation. Remember what I stated earlier in this book about representing your department? This is exactly why you engage. She asked me, with her finger pointed at my face, Why were there so many firetrucks here? Why did we send so many firetrucks and firefighters to a simple food-on-the-stove call? She stated she thought it was a waste of money and resources, and we should be able to handle it with less. After all, there was no fire. Once she was done asking, I started to give her my answer. First, I introduced myself and told her my rank. Then I began telling her that our emergency response was derived from department policy which dictates how we respond and with what apparatus to each type of call. I explained what each apparatus role is in the assignment. Then I told her about the National Fire Protection Association (NFPA) standards for response, and how redundancy is the key to a safe and successful outcome at fire incidents.[7] I told her about the fact that we really don't know what is happening until we show up, and if there is an actual fire, we need our resources there now, instead of having to call for them and wait. Then I explained how our city taxpayers enjoy a lower rate on their homeowner's insurance because we are an Insurance Services Office Class 1 department based on how we operate.[8] Basically, I laid it all out for her to fully understand why we sent the fleet. She smiled and thanked me for my time and for explaining to her why we do what we do. As she was walking away, I told her my name again and what firehouse I worked at, and I asked her for her contact information in the event I could assist her in any other way. She thanked me again and walked away. When I returned to the group of firefighters packing the hose, someone asked me what that lady wanted, and I explained what had just transpired. Someone else asked if she was going to turn me in for something I may have said to her. I stated no and explained exactly how I responded to her. Then another firefighter asked if I had gotten her name. I said no, but I already knew who she was. This lady was a state legislator for our city who happened to live in the neighborhood where this incident took place. I had recognized her from seeing her at various political events and on TV. But for me it didn't matter who she was. I was going to deliver the very same oral-board-worthy answer to anyone, whether it was a politician or a woman on the street. I could have told her to buzz off, but this situation presented an opportunity for both of us. The point is, I gave a complete, comprehensive, definitive response to her so I could practice for my upcoming oral exam, and also because I was representing my department and myself as the officer.

The fourth takeaway is to remember that as the officer or aspiring officer, you are the designated problem-solver. You are the person who the crew comes to for answers, direction, and leadership (fig. 3–6). So do the people serving above your rank, like chiefs and administrators. Likewise, the public comes to you for answers as well. They want to speak to whoever is in charge, not the hydrant firefighter or the roof firefighter! You're it. You need to act accordingly and be that problem-solver. This is why it's so important to not only know your job thoroughly, but to know your resources thoroughly as well. So, when you walk into your oral exam, remember your primary function is the problem-solver,

---

7. *NFPA 1710: Standard for the Organization and Deployment of Fire Suppression Operations, Emergency Medical Operations, and Special Operations to the Public by Career Fire Departments* (Quincy, MA: NFPA, 2020).

8. Hartford Fire Department, *Administrative Manual—Department Directives*, 2015.

FIG. 3–6. Pictured here is Firefighter Alberto Vazquez from my engine company crew. We are about to take a ride downtown to fire headquarters so he can get promoted to driver (pump operator). Certainly, a proud moment for both of us, for him receiving his promotion, and for me, mentoring, guiding, and prepping him for the exam to a successful and fruitful conclusion. That's my job as his officer.

period! If they throw a fireground scenario at you, solve the problem as best you can in the specific job function you are testing for. If they throw a firehouse or public interaction scenario at you, solve the problem. Listen carefully through all the buildup and chatter, find the heart of the problem they are asking you to resolve, and simply do it! It's easy to get flustered and start to freak out, mostly because you are nervous and probably inexperienced at taking exams. *Don't!* You are the boss here, not them. They are asking you for help to solve this issue, so do it. Think to yourself, "What if I was at an incident or in a firehouse and this happened to me; what would I really do here?" You have no one to turn to for help. You must know what to do and what resources you need to solve this problem. Just remember to be thorough in your response; don't leave anything out or omit steps. If you think you did, most often you can go back and fill in the blanks at the end of the oral exam; when they ask, "Do you have anything to add?," that's the time to go back and fill in anything you think you may have left out. If you prepared well enough, you probably didn't miss anything, but you simply don't remember if you said this or that. That's a normal human reaction. Bear in mind; the panel is asking you to problem-solve, so be the problem-solver.

The fifth takeaway is a personal one. I learned the hard way that I needed to do this because I would become overwhelmed with anxiety and emotion after each test, whether it was the written or the oral. So here is what I do: once you complete the exam and you walk out of the building, get in your vehicle and go home. If you are returning to your working shift, go hide somewhere in the firehouse. You need to decompress. You have worked yourself up mentally to remember all the information and put your mindset into the exam mode. Now is the time to decompress and not think about any of this stuff for a while. You took the test and did the best you could, now leave it alone for another day. When I was asked how I think I did, I simply said, "We'll see!" The problem arises when you start to commiserate with other firefighters who took the test with you. The old "What did you say for this question?" or "What did you put for that question?" will drive you nuts and make you more anxious and nervous. When I took the captain's exam and returned to the firehouse to finish my shift, my phone was ringing off the hook with guys who I had just taken the test with, peppering me with what my responses were to the oral exam. To be honest, I started freaking out because I didn't even remember being asked that question, and my mind started to confirm that indeed I had flunked this test. This feeling is only compounding an already anxious and tenuous day; why make it worse? Unless you thrive on mental anguish, simply disengage the testing process once you are done, and give yourself some time to decompress. You can engage in the question-and-answer session another time when you're ready.

Case Study

In 2006, I tested for the promotion opportunity for deputy fire chief. This was not only the logical succession of my career path from captain to deputy chief, but one which I had been preparing for both in study and in exercise since being promoted to captain in August of 2000. I studied hard for the written and answered as best I knew how at the oral exam, but out of 15 eligible candidates who successfully made it to the promotional list, I was ranked number 12. I didn't do as well as I had hoped, and the odds were stacked against me with seniority and the other captains who simply scored better than I did. In the end, only 8 of the 15 were promoted, and when the list expired, I was without a promotion. When I received my ranking a few weeks after the exam process, simply put, I was devastated! I fully knew there were only going to be a few promotions and not the entire list. My whole world had turned upside down and I believed I had failed, both as an officer and as a man. I was clearly emotionally depressed, and it took some time for me to grasp the concept of failure and how to grow from it. My family played a big part of my rebound success in that naturally they stood by me even though they all thought I was wallowing in self-pity, and they clearly held me in high regard in all factors of my career. There were two individuals who, unbeknownst to them, had the greatest impact on my mental turnaround and ultimately the path to success for the next promotional opportunity for chief officer. One was a seasoned, salty deputy chief I encountered one day during my normal course of work. We began talking of the exam and how I unsuccessfully fared, and he began to lay out all the successes he has seen me achieve, and how my attitude, disposition, and work ethic bled into everything I am involved with, both professionally and personally. He went on to boost me up and explain that failure is necessary for professional and personal growth, and to use my newfound study time wisely to not only prepare for the next exam, but to learn the job of chief officer intimately and embrace the core duty, values, and vision of a chief officer. This was to me both profound and enlightening, for I had never thought of this promotional endeavor and the achievement of the chiefs' rank as anything like this before. I left that conversation reinvigorated and full of energy and steam, and I set my sights on getting that next promotion the right way: not just scoring high on the exam, but by being prepared and ready to take on the newfound responsibilities and duties of a deputy chief. The other individual was one of the newly promoted deputy chiefs from this exam that I just took where I did not receive the promotion. We were dispatched to a local hospital for a reported bleach spill inside the building where the hospital laundry is cleaned, and this new chief was the incident commander at this incident. You could tell in his demeanor and voice that he was clearly nervous and trepid as the new chief in charge, even though he had several more years on the job than I did at the time. He was going through his assignment tasks with the companies on scene, giving orders and direction and communicating with his company officers. My company was third due at this incident, so we walked up to the command post, and I asked him what he wanted us to do. Clearly tense and seemingly anxious, he quietly turned to me and said, "You stay right here by my side, I need you with me right now!" I was totally floored. I had known and worked with this chief when he was a company officer, but I didn't expect this request to come from him now, let alone for him to ask me, the guy who couldn't score high enough to get promoted. I stood behind him and assisted in processing the incoming information he was receiving and made minor suggestions as to how to proceed until the incident was mitigated. He was relieved and thanked me profusely. Afterward, I asked him why he grabbed me to help him, and he simply stated that a promotional exam is not the truest gauge to determine competence and ability; it's experience and the fortitude to deliver. From that point forward, I held my head high and embraced my role as senior captain on shift because this was a definitive point of empowerment for me personally and professionally. I was galvanized!

Questions

1. As we all know, a successful promotion is based mainly on attitude and empowerment, rather than simply passing the exam. How would you go about empowering, enlightening, inspiring, and uplifting those individuals who came to you for assistance in the endeavor of promotional testing?
2. How would you convey the identity of the go-to mentor to those seeking direction and guidance, or to those who are lacking the inspiration and motivation to go the extra step or to simply do better at their jobs?

4

Decision-Making: Know Your Job!

Obviously, to be effective at your duties as an officer, you must be intimately familiar with the parameters, authority, and responsibility of the job you hold, whether you are a lieutenant, captain, or chief officer of some type. What are the parameters of your job? What is your official function, who do you report to, who or what do you lead, and what do you do on a daily basis? What are your boundaries? How do you fit into the department in terms of your present job duties? You know your parameters when you are tasked with doing something that you know is not your job, right? Do you know the parameters when you are put in a situation that you're not trained for, not prepared for, and have no experience in, such as a lieutenant ordered to do something that an administrator would normally handle? What exactly is your authority as an officer in your present capacity, and how are the other ranks above and maybe below you different? What changes with their authority? Does it increase and become broader? Does the scope change? How about your responsibility? What exactly are you responsible for, who or what are you responsible for, and what happens when you don't meet those responsibilities? All great questions, but you should know the answers before you get that officer's badge pinned on your uniform (fig. 4–1). These questions are based on experiences I've had when someone told me, "Stay in your lane" or "That's not your job," or when I omitted something I should have done only to find out that indeed it was my job to handle. Also, if you are a career officer or in a unionized department, you should know the details of your collective bargaining agreement (CBA), also known as your contract; however, many departments do not have a CBA due to being in a right-to-work state. Therefore, your department policies take precedent and are equally important. Within that, there are numerous details on what your role, responsibilities, and authority are. You'll make friends really fast the minute you violate the contract: your phone will blow up and someone will start screaming at you. You may have gone above and beyond the parameters of the contract, and your behavior brings swift scrutiny and

FIG. 4–1. As assistant chief, I would respond to working fires and confer with the incident commander, then offer any assistance that may be of use. (Photo courtesy of Pat Dooley)

possible discipline. So instead of always getting yourself in a jackpot with the union and administrative leadership, take the time to study the contract, and keep a copy handy for reference. Within your job description there is a narrative that spells out what exactly your duties and responsibilities are. The job posting for an officer usually has this narrative listed on it, and it derives from the agreement between the labor union and the human resources department of your jurisdiction. There is really no room for interpretation—it's all spelled out.

For me it was simple: what is the job of (line) fire lieutenant? The job description states you must[1]

- Be responsible for responding to all calls for emergency service such as fire, emergency medical services (EMS), rescue, hazmat, and other related calls for service and document fully.

---

1. City of Hartford Human Resources Department, "Fire Lieutenant," https://www.governmentjobs.com/careers/hartfordct/jobs/newprint/2102562.

- Conduct and attend training classes and evolutions including classroom and hands-on training.
- Conduct fire prevention duties within your assigned district, including public education elements, building and neighborhood surveys, and submitting reports.
- Conduct and participate in equipment and apparatus maintenance, firehouse and property maintenance, and associated duties.

One of the most often overlooked and misunderstood core duties of an officer is that of safety. Safety is an attitude, not an option. Clever does not always equal smart. The Swiss cheese model of human error causation from *Fire Engineering*[2] is a perfect example of how to explain fireground and other failures that can cause injury and death. As the failure bomb worms its way through the holes in the block of cheese, if it is successful in passing through all the various barriers presented, then when it comes out the other side it will surely explode and cause harm (fig. 4–2).

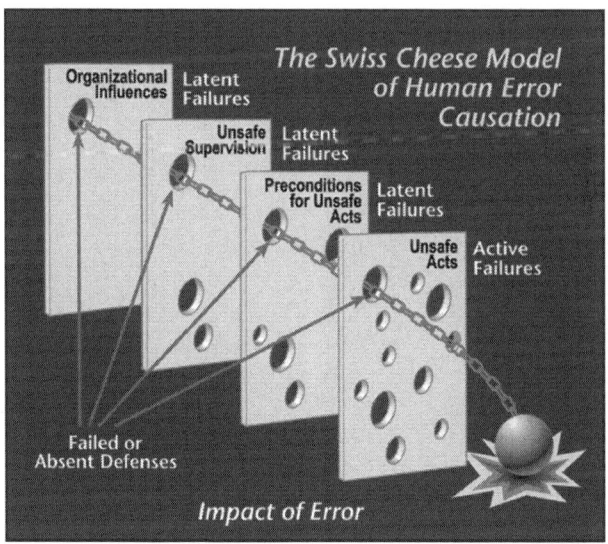

FIG. 4–2. I often cite this Swiss cheese model for failures as an illustration in understanding why bad things happen. (Illustration courtesy of *Fire Engineering*)

The model points to failed or absent defenses, organizational influences, unsafe supervision, preconditions for unsafe acts, and then the actual unsafe

---

2. Swiss Cheese Model, in Quinn MacLeod, "Managing Fireground Errors," *Firefighting* (blog), Fire Engineering, March 1, 2010, https://www.fireengineering.com/firefighting/managing-fireground-errors/#gref.

act. You as the officer are one of these defenses, so actively being cognizant of this role and the impact you may have on the prevention of a catastrophic failure cannot be overstated. By virtue of your capacity, you become the de facto safety officer both on the fireground and in the firehouse. It is incumbent upon you to take control when you see things that are functionally not right and could or will lead to a safety issue, mainly somebody getting hurt or worse. I will give one subtle example to make my point. This is not to say that I'm the hero here and everyone else at this incident was compromised. My goal is to present an example of an incident falling apart for whatever reason, and someone stepping in and making corrections before it's too late. In April 2015, I was operating as the safety officer at a fire in a two-and-a-half story wood frame balloon construction home in the city's north end (fig. 4–3).[3]

FIG. 4–3. During suppression evolutions, the fire grew in intensity within the walls of this house, compelling me to take immediate action as the safety officer. (Photo courtesy of Pat Dooley)

The fire was initially knocked down and companies were basically opening up and hitting the hotspots. As this overhaul evolution progressed, companies started to inform the incident commander (IC) that more assistance and attention was required in the central core of the structure on both the first and second floors because the fire had worked its way into the walls and began growing and traveling upwards. At the same time, for some unknown reason, a pump operator inadvertently shut down a handline that was being utilized to stop the spread of fire on the second floor. As radio reports of conditions and needs started to bombard the IC, he unfortunately became confused, lost track of what company was where and what they were doing and began making decisions and giving orders that were inappropriate and unnecessary, in the belief that he was in fact correcting the situation.

---

3. Hartford Fire Department, 68 Westbourne Parkway (1-1-2), April 10, 2015.

My function at this fire was that of safety officer, and I immediately recognized that this situation was quickly getting out of control and could lead to something bad happening (fig. 4–4). While overseeing interior operations, I immediately got on the radio and began correcting the companies as to what they were supposed to be doing and where they were supposed to be within the building. Then I notified the IC that he also needed to manage certain resources more carefully. As the work progressed, the companies were able to gain control of the situation and the risk was mitigated. Ultimately the fire was knocked down and the incident safely stabilized (fig. 4–5).

The point being made here is basic: take control for safety's sake! Do not let the situation get out of hand until something catastrophic happens, and you stand there and say you knew this was going to happen. Why didn't you try to stop it?! My role as safety officer dictates intervention when needed. Also, as senior deputy chief and tour commander, I have an obligation to step in when I believe I need to, regardless of whose feelings get hurt or what egos get bruised. I cannot stand by, sitting on my hands, and watch an event unfold that may lead to someone getting hurt or killed. I felt compelled to act at this incident, and when need be, I have done it at others. The IC and the companies were losing control so all I did was intervene to right the ship. Once things were back on track, I went about my original function as safety officer.

FIG. 4–4. Managing resources properly is important to safety and success. (Photo courtesy of Pat Dooley)

74 An Insider's Guide to Mentoring the Fire Officer

FIG. 4–5. A safety officer has an obligation to intervene before a catastrophe can happen. (Photo courtesy of Pat Dooley)

Indecision Is No Decision, So Make a Decision

As the officer or person in charge of a company or crew, your primary function is leadership and management, and by design, decisions will need to be made by you. If you believe you are not up to the task of decisive leadership, you surely will have problems in your role as the officer. It's one thing to not have enough information to make an informed decision. What I am referring to is the human element of failure to lead and decide what is best. Our decisions need to be definitive and decisive. You can often go back and amend those decisions based on new, relevant information, but at least you have already decided to act, intervene, or basically to proceed. When you decide, based on your training, education, and experience, what needs to be done, you should stand behind your decision and defend it. Don't run or cower from it; you are in charge and that's why you are there: to decide and act. Decisions should be made with sound judgment and good reason, not simply flippant or uninformed. Earlier in this book I talked about the scientific method, and that's the formula I use to engage critical thinking skills. One glaring example comes to mind that will explain what I mean.

I was the overtime company captain in charge of our rescue unit one Saturday in March 2006, which just happened to be the day Hartford held its annual

St. Patrick's Day parade. I was informed by the tour commander in the morning that my apparatus would be in the parade with our color guard along with those members marching from our department. He stated if something big came in, meaning a serious call requiring our rescue apparatus, we were to just leave the parade and respond accordingly. When the parade began around 10:00 a.m., we slipped into the marching line behind the group of firefighters that were representing our department. As we weaved our way through the downtown streets in the parade route, we noticed the unusually high number of crowds and revelers watching the parade. After a few minutes of being in the parade, our apparatus radio started blaring a call for service from dispatch reporting a building fire in the west end of the city. I looked over at my driver and told him to continue in the parade, and that we would wait for the first-due apparatus's size-up radio report to determine if we would in fact leave the parade, because often, calls for reports of a fire are erroneous or misidentified, and I did not want to totally disrupt the parade for an insignificant call. Besides, these kinds of things always happen at the wrong time, whether you're on the toilet or in the shower, just sitting down for dinner, or even in a parade. However, the first arriving company on scene reported a working fire. I looked at my driver and said that we needed to find an opening and get out of this parade immediately (fig. 4–6).

Each intersection which would normally provide us with an easy escape route from the parade was clogged with people watching the parade. We approached at least five different intersections, and each one was unusable because it would require the people to part ways and make a hole big enough to drive the apparatus through, and then we would still have to move the wooden barricades blocking the roadway into a safe position. One of the larger intersections we came upon to find our exit was blocked by a man in an electric

FIG. 4–6. Screenshot of a video showing the tactical (rescue) unit responding through the parade route to a two-alarm working fire. The driver of this apparatus received injuries at this fire which required medical attention and ultimately time off to recover. (Screenshot of video clip courtesy of TRFIRE86)

wheelchair, and he had difficulty trying to maneuver out of our way. We were wasting time by now; I had already acknowledged to the dispatcher that we were en route to the fire. Because the Hartford contingent was at the front of the parade and nobody else needed to move out of our way, I made the decision to continue through the parade route at a safe speed and exit through the end of the route and then continue on with our response. So, we proceeded with our warning devices activated (lights and sirens) to continue through the route. We finally came out at the end and then proceeded to the working fire, which by now had been escalated to a second alarm. We arrived a few minutes later and performed our assigned functions. During the overhaul phase of the incident, my driver was pulling ceilings with a Hartford Plaster Hook[4] and was splashed with hot roofing tar on his face near his eyes. He was subsequently transported to the hospital and put off the line to recover.

When we arrived back at quarters, the tour commander summoned me to his office. He stated that there was a police officer directing traffic in the parade who made a complaint about what transpired with us responding. He stated the officer complained we were speeding, should not have responded, and that we should all be fired and arrested. This was the typical antagonizing friction we were used to from our police. When the tour commander got done telling me there was a complaint, not only from a police officer but from several members of the public who were perplexed as to why we left the parade, I began my response. I explained that we tried desperately to exit the parade to no avail, that we in fact did not speed and could prove it by downloading the data recorder in the apparatus (most notably used if there was a serious crash involving the apparatus), that this was a real fire with the lives of civilians and firefighters at stake and not some smells and bells alarm, and in fact my driver was injured and subsequently put off the line fighting this fire. I went on to say that I decided as the officer to proceed carefully through the route, that I stood behind my decision, and that I made it with sound judgment and definitive reasoning, and that's why I was here, to make decisions as part of my job. The tour commander stated he would have done the same thing because there really was no other option. No further investigation or discipline was warranted. I made a decision as the officer and that's where this entire problem ended.

Part of making proper decisions is possessing the knowledge, skills, and abilities to know your job. Many times, when officers or acting officers roll up to an incident, they have an idea of what they should be doing, but quickly become overwhelmed and distracted by what's happening. I have always relied on the three incident action templates by priority to help guide me, so I don't become confused, and more importantly, I don't miss anything. I do not claim

---

4. "Hartford Plaster Hook," Capital City Industries LLC, 2021.

to have invented them; their origins are found in various training manuals. I merely utilize these existing templates to my advantage as they were intended by their authors. They are simplistic, basic, and fundamental, and I swear by them for all incidents, and I think you may benefit from them as well.

For a fire incident, the template is *rescue, exposures, confinement, extinguish, overhaul, ventilation,* and *salvage* (RECEO VS).[5] For this algorithmic template, the order may be jumbled or have elements omitted to adapt to the actual fire incident, but the basic template is spelled out, so you don't miss anything. Many times, I have heard new officers and persons in charge ask, "Where do I start?," and I always answer, "At the beginning!" (fig. 4–7). RECEO VS has been around for years, it's tried and true, and I used it every time I oversaw a fire incident.

FIG. 4–7. Rescue is the first priority in RECEO VS. (Illustration courtesy of clipart-library.com)

For a hazmat incident, we have all been trained to utilize the eight-step process, and that works well every time without having to rearrange the sequence of the steps (fig. 4–8). The eight steps are *site, identification, assess, clothing, resources, control, decontaminate,* and *terminate*.[6] It works every time and is thorough and reliable.

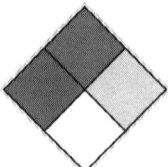

FIG. 4–8. The symbol for hazmat incidents at fixed facilities (*National Fire Protection Association [NFPA] 704: Standard System for the Identification of the Hazards of Materials for Emergency Response* placard). (Illustration courtesy of clipart-library.com)

---

5. International Fire Service Training Association (IFSTA), *Chief Officer*, 2nd ed. (Fire Protection Publications, 2004).

6. Gregory Noll, Michael Hildebrand, and James Yvorra, *Hazardous Materials: Managing the Incident*, 3rd ed. (IFSTA, 2005).

For an EMS call, the template is by priority of things that could kill someone: *airway, breathing, circulation, bleeding*, and *shock* (fig. 4–9).[7] No need to reinvent the wheel here; this works every time. When you arrive at an EMS call, just apply this template to your patient and you will be fine. Like I say, do this and everything else is gravy! These templates address the meat and potatoes of the incident.

FIG. 4–9. Universal symbol for EMS (Illustration courtesy of clipart-library.com)

Working with a Crew

As an officer or person in charge, you will most likely be working with a crew or company, or several companies. Teamwork is essential for success, and you need to build that team up so they may work together and in the same direction for the same result. I have always considered myself a pragmatist, in that I look for the best info and ideas to make an informed decision when I can. I don't believe I hold all the answers to everything, which is why I enjoy collaborating whenever possible. This does not mean I relinquish my responsibility or diminish my authority; it simply means I seek the best possible input before dropping the hammer. Not every decision is managed this way. I also believe that the firehouse can appear more like a democracy where input is welcomed. But on the fireground or at an incident, I hold the authoritative position of leader, taking in only necessary input to make an informed decision. I do not engage in debate at an incident. It is wise for the officer to recognize the experience of others in the crew and be open to suggestions when making decisions, because those with prior experience can assist in properly influencing your tasks and ultimately accomplishing your goals. When your crew works as a team and completes a task that you believe is favorable, don't be shy about praising your crew or individuals in public, especially to your bosses. However, if one of them screws up, or the entire crew makes an error, give them the talk in private, away from their peers and the public. Respect their dignity! Correct them, if need be, and show them how to make this correction for the future. I have heard this saying time and again in my career: it's not how you screw up that's as important as how you will recover and keep it from happening again. Obviously, if the issue is any way related to safety, it needs to be addressed immediately, regardless of the surroundings or circumstances. Safety takes priority.

---

7. American Academy of Orthopaedic Surgeons, *Emergency Care and Transportation of the Sick and Injured*, 11th ed. (Jones & Bartlett Learning, 2017).

MILFORD STREET FIRE

This April 16, 1997, two-alarm fire at 34 Milford Street was one of three fires I responded to while assigned to Ladder Company 4 in the city's north end which almost ended in my line-of-duty death (LODD). This fire was in a three-story wood frame balloon legacy construction, and upon arrival we encountered all three floors of the wooden rear porches ablaze. Our dual engine and ladder company quarters was right down the street, and when we arrived with the accompanying engine company, the engine officer informed me he would be flanking the rear porches with a 2½" attack line to knock the wind out of the main body of fire, thus providing me and my crew with an opportunity to enter and perform a primary search. My crew and I made our way to floor number two and performed the standard search, rescue, vent evolution. Believing the exterior attack line was in operation, my crew and I searched the second floor and took out the windows in the kitchen simultaneously. The kitchen was in the rear of the building and the fire had started to burn through the door separating the kitchen from the rear porches. From experience I knew I had about a minute or two before it burned through before we had to bail out, but by then the attack line would have alleviated this issue. One of my crew stayed in the dining room next to the kitchen and kept an eye on conditions and to provide guidance if a bailout was required, while myself and my other crew member searched the two back bedrooms and the bathroom. The crew member with me finished his search and returned to the dining room where the waiting crew member was, thinking I was right behind him. Unfortunately, when I appeared from the bathroom having completed the search, thankfully finding no one, the rear door in the kitchen was burned through and the kitchen was now filled with fire. I had to make a quick decision: abandon my crew and bail out of a window on the second floor, hopefully by ladder, or scurry on my stomach as fast as possible through the kitchen with the fire overhead to rejoin my awaiting crew in the dining room next to the burning kitchen. I decided to go back the way I came and scurried through the kitchen, thankfully making it to my awaiting crew members. When I appeared through the fire filled doorway between the kitchen and the dining room, they were shocked to see that anyone could have survived that maneuver. By now, a line was being brought up the front stairway and put into action by the second-due engine company. The first-due engine's large attack line, originally promised as being engaged in a rear porches exterior flanking maneuver, was somehow swapped for a smaller line and diverted to the first floor. This was an unfortunate turn of events, but one heck of a story. The purpose of this story is that when working with your crew, you all instinctively become familiar over time with how each of you operate and the expectations set forth by the company officer, and often synergize into a cohesive team to effectively engage your tasks. Although each of you is an individual, the sum of the parts makes the whole, and in this case the crew. The photo seen in figure 4–10 was taken after the fire was knocked down and we were picking up. The two other fires which I almost met my demise were a building collapse and a tenement fire where again we were doing a search, and when we finished, we subsequently were trapped by heavy fire blocking our exit. Stories for another time. It's these close calls and quick decisions that illuminate the gravity of what firefighters experience all too often. These three near-death incidents for me, all within an 18-month timespan while assigned to Ladder Company 4, provided context and perspective to what I thought I was an already seasoned firefighter. Me and my crew placed greater emphasis on our safety and survival skills.

FIG. 4–10. Milford Street fire, Ladder 4 Group 4 (Photo courtesy of Joe Marino)

Radio Communications

Communicating on the radio system is a necessary function of our duties as firefighters. But for some, however, it can be elevated to a sport, and even seen as entertainment for the masses. It is a tool that should be used carefully and sparingly. If you're at a working fire or an incident and you're yapping incessantly on the radio, you are simply not working. Instead of a tool in your hand, you have a microphone, and you are engaged in verbal communication instead of an evolution or a task. This is not to be confused with necessary and timely radio messages or alerts about a safety issue. If you are tying up the radio with your broadcasts, someone who may in fact need the airtime to report a serious safety condition or worse, a Mayday, may need it more than you. Please don't be that person. Keep the airway clear for a possible Mayday call.

Also, when you do talk on the radio, be sure to provide clear, coherent, concise quick reports. Don't provide a dissertation on the details of what's wrong with the building during your size-up, such as "the paint is peeling" or "the lawn hasn't been cut for a week," or go on for several minutes when communicating with other companies or the dispatcher. Less is more! We have all experienced dealing with that one person who clogs up the radio every time they go out the door. During my career, we had several officers who, when they got on the radio at an incident, communicated so poorly that those of us not at the incident but in the firehouse would mute the television so we could hear the radio transmissions. They became entertainment for us because of the predictable embellishments and often combative tone used when speaking. I cannot tell you what exactly you should state while speaking on the radio—that must derive from your department policies and your training. I am only illuminating what you should not engage in when transmitting. When arriving first due or dispatched alone as a single company to an incident, as tour commander I would expect my officers and persons in charge to state just the basics in their size-up report. I instructed them to paint me a picture, not describe. I need to visualize what you are seeing so I can make informed decisions on my end if need be. Tell me what you have, what you are doing, and what you need (fig. 4–11).

For example, I had been monitoring the radio traffic of a young firefighter who was acting in the capacity of company officer for the first time on tour. As the day went on and she responded to more incidents, I noticed her radio reports were lacking basic information needed for everyone listening to understand what was going on. I showed up to a motor vehicle accident her company was dispatched to and pulled her aside once the call was complete. I told her I understood how overwhelming and confusing it can be to give radio reports,

FIG. 4–11. This car fire in the rear of a junkyard looked like a five-alarm blaze while responding to it, and several neighbors called in a huge fire at this location. Upon arrival, it was a stack of junk cars burning nowhere near the building. My radio report was simple: what do I have, what am I doing, what do I need? Now everyone else responding to this incident knows exactly what's going on.

but this template is what she needs to use to avoid complications: What do you have, what are you doing, what do you need? That's it. On her next dispatch for a motor vehicle accident, her radio report went as follows: "Engine 10 we are on-scene of a two-car motor vehicle accident, we are command and are checking for injuries and fluids on the ground, keep the remainder of the assignment responding until we confirm," followed by, "Engine 10 to dispatch, we have no injuries or fluids, we can handle this incident without the remainder of the assignment." That's it—simple, basic, and quick. She didn't take 5 minutes to spit out a complex, winding, and ambiguous message. If there needs to be something pertinent added, then of course state it, otherwise, less is more here.

Committing Resources Unnecessarily

As an IC stands in front of a burning building, it's easy to get overwhelmed with decisions. This is true as the company officer or person in charge, but what about the firefighter who is deciding where to throw ground ladders or place the bucket of a tower?

Look carefully at this photo seen in figure 4–12 of an incident I was the IC for. When the decision is made about where to throw ground ladders or to place a bucket of a tower, think of those people in the most imminent danger, not those making the most noise. As you can see, this dude is in the window indicating that he is prepared to jump. Most often, people panic because they hear alarms and smell smoke during a fire, but as anyone who has been to a working fire knows, people don't always think rationally in an emergency. This individual does not need rescuing; he needs to stop panicking and wait for the

4 • Decision-Making: Know Your Job! 83

FIG. 4–12. Look at what is happening here as opposed to what you have been swept into thinking is happening due to the inherent chaos of the fireground. (Photo courtesy of Pat Dooley)

help to assist him exiting either via the window, which is always inherently dangerous, or simply by being guided down the interior stairwell. Some modern tenement construction has smoke dampers to evacuate smoke from the egress stairwells. If not, then mechanical ventilation is required from fire companies. Communicate to this guy that help is on the way, but your bucket or ground ladder should first be directed to those in imminent peril, such as those closest to the actual fire or firefighters engaged in suppression evolutions who may need to make a hasty exit. This guy needs to stop panicking, and so do you! If you placed the bucket to this guy and committed this valuable resource to someone who obviously has no smoke billowing overhead, is not reacting to a high heat condition, and just thinks it's time to bail out, someone closer to the actual fire, heat, and smoke who probably needs the rescue more that this guy might be hurt or worse. Once those compromised individuals are mitigated, then go after the easy plucks, as we used to say. The ability of mental strength under duress is a skill that needs to be honed for both firefighters and officers.

Roof Ventilation

This photo seen in figure 4–13 of my neighbor's house is a solid illustration of the challenges the fire service faces every day. More and more homes and businesses are installing these solar panels on roofs, and thus have created a huge challenge for arriving companies during a working fire.[8] These panels cover the entire working area of the roof on both the front and rear of this house. They clearly impede rooftop ventilation for us. Although they are mostly lightweight

---

8. Jeff Simpson, "Solar Power 101 and the Fire Service," *Firefighter Training* (blog), Fire Engineering, October 17, 2016, https://www.fireengineering.com/firefighting/solar-power-firefighting/#gref.

FIG. 4–13. Solar panels present a new challenge for firefighting.

and don't add too much load to the roof structure, keep in mind two thoughts. First, these solar panel companies will sell their product to anyone. Sales first, and then maybe worry about a potential emergency. I have seen panels installed facing north on houses, and that is the most indirect route for sunlight. Second, although the salespeople will tell you that your house can handle the additional deadload, do they really know the state of structural integrity of your roof? I'm not slamming the industry, just the installers who are under the direction to put them up. These panels maintain a constant electrical charge from the panel to the meter, and some households may also have a battery backup or storage system, which can hold even more energy. Panels still collect energy during daylight hours, even when it is cloudy and overcast, whether they are electrically connected to a grid or meter, or not. This presents an ever-present danger to firefighters working around these installations. Moreover, battery storage and backup systems connected to the solar array may be present in the structure as well, which again increases the level of danger for suppression crews. The problem also is when you go up to ventilate during a fire. Where do you put the hole? How do you get that hole made with these panels in the way? Do you start crashing and bashing these things, or try to unbolt them? How slippery do you think they are being smooth panels? Add ice, snow, and rain, and then what happens? My department has already responded to fires caused by critters chewing on the wiring and connections. This is a headache the fire service doesn't need. As the IC giving the order for topside ventilation, you can expect a delay in completion due to these panels and may need to alter your tactics, or worse, your entire strategy because of this impediment.

Offensive or Defensive?

As an officer in charge of this working fire, the decision to place companies inside is a difficult one. Not including the life-safety issue and rescue, if this building was unoccupied, would you commit personnel to the interior to perform suppression evolutions? What are your criteria for a go or no-go decision? Is it safe for crews inside, all things considered? What is the risk versus reward? This is the decision that needs to be made constantly, repeatedly, and frequently when companies are operating inside. As the IC, what am I gaining from this, and does it change the outcome of the scenario? Can I be just as effective from the outside, without all the risk? God forbid, what if there was a structural failure and someone got trapped, or worse, ended up dead? For what?! Did we save a structure that, according to the fire marshal and the building inspector, would be knocked down with a backhoe at first light? How would you explain a LODD when you were questioned about your decision to put companies in this building? Please don't misunderstand me here; I am not the hit-it-hard-from-the-yard type of IC. What I am saying is, as the IC, you need to make a sober and cogent judgment of what is the risk versus what is the reward for making that decision. One time at a large working fire, I was the IC standing by the command post conferring with the other chief officers about the status of the fire. I did not notice the mayor standing behind me, but he clearly got an education on how we do our job that evening. A chief standing near me asked if I was going to allow companies inside, after this fire had been burning for a while and clearly the structure was compromised. I made the off the cuff remark, "I'll burn the whole city down before I kill one of my guys!" to him. The mayor was not happy when he heard my utterance, but the precarious situations we are placed in often demand life and death decisions. As the IC, it's my fire and I will not intentionally make a rash decision that could result in a LODD! Risk a little to save a little, risk a lot to save a lot! That's what has always been drilled into our heads (fig. 4–14).

This photo in figure 4–15 illustrates an example of what can happen when the plan does not go according to schedule. We had three towers pounding the snot out of this two-and-a-half story, wood frame, balloon construction fire. Sounds effective, but things started to go bad. Each bucket had two master streams and they at times were overshooting the fire building and tearing up the exposure buildings, the rear garage, and the driveway. One firefighter received a facial injury from flying asphalt. Remember, we are not here to make things worse. Along with the three towers operating, there was a significant water supply issue. There were only enough hydrants to feed two towers, so an engine company had to stretch a supply line down a neighboring driveway, through the rear yard, and into the yard of the next street over to grab a hydrant

FIG. 4–14. This figure shows companies operating in a defensive posture after being chased out of this two-and-a-half story, wood, balloon construction job. Knock the main body of fire down, let the water drain, then go in and assess. If it appears structurally compromised from the outside, stay out. These types of fires are common in Hartford and the Northeast. (Photo courtesy of Pat Dooley)

FIG. 4–15. Three towers pounding this wood frame working fire with their master streams (Photo courtesy of Pat Dooley)

on the next street. Very creative. Another issue arose because upon arrival, one ladder company jumped into the bucket of another ladder company and began to operate. Then the first ladder company jumped into the bucket of the second ladder company and began to operate. This made sense because the first-due ladder company entered the structure to do a search, rescue, and vent evolution. When the second ladder company arrived, the order was given to go defensive, so they jumped into the first ladder company's bucket because it was already in position. Unfortunately, these company officers did not communicate this to me, and during the fire I was directing these ladder companies by their assigned designation, and not by where they actually were. When I directed Ladder 4 to move to a certain angle on the B side, and they were actually on the D side in the bucket of Ladder 3, this caused a lot of confusion, and those company officers got an earful from me after the fire. The point? If changes need to be made at the company level, they obviously need to be clearly communicated to the IC. Companies don't need to add to the confusion and chaos by making decisions that they don't realize have a profound impact on the command operation.

Rapid Intervention Teams

In Hartford, our assignment for a confirmed working fire or other labor-intensive incidents automatically designates an engine company as our rapid intervention team, or RIT team as they are referred to.[9] Once the assigned company arrives, they gather their RIT team equipment contained on our heavy rescue unit and report to the command post to await further orders. Often, they just stand around and buff the operation rather than be proactive in their mindset and approach. What they should be doing is preparing for the inevitability that they may be placed into operation to rescue their fellow firefighters (fig. 4–16). The RIT team should have the mindset that although they are on scene of a fire and standing by as rapid intervention, they should be preparing for a rescue situation. Treat it like you are at your own incident within this initial incident. That way, when something does go down, you and your crew have already been preparing! If nothing bad happens and you're not needed, then it was a good practice session. Even though the rescue company checks the gear out each morning in their daily routine, the assigned RIT team might not have seen this equipment for months due to the frequency of being assigned as the RIT team. The assigned company should be checking the equipment and starting the saw, as well as testing out all the other RIT team equipment for readiness. I used to direct my RIT teams to throw ground ladders to the fire building, specifically where interior companies were operating. As long as I can redirect the RIT team in a pinch, I can use them to throw ladders[10] because those ground ladders are to facilitate the rescue of companies operating inside the structure. Basically, initial ground ladders thrown in the first few minutes of a fire are to rescue civilians who may be trapped. After that, those ladders are in place for us, the firefighters, to facilitate rapid egress in an emergency or Mayday situation (fig. 4–17). It is our department policy to remove security bars and screens on windows to facilitate rapid exit as well. That task is automatically assigned to

---

9. Hartford Fire Department, *Administrative Manual—Department Directives*, 2015.
10. *NFPA 1500: Standard on Fire Department Occupational Safety, Health, and Wellness Program* 8.8.8(2) (Quincy, MA: NFPA, 2021).

FIG. 4–16. The RIT team standing by at a working fire (Photo courtesy of Pat Dooley)

an additional ladder company; however, while that truck company is en route, the RIT team should also be initiating this evolution.

As the IC, I always told my company officers when referring to the order to deploy ground ladders, "Keep throwing them until I tell you to stop or you run out of them!" Why? Who are these ladders for? Us! They are put in place to rescue us firefighters. Also, as the IC, once the ladders are in place you need to notify the companies operating inside that in fact ground ladders are in place if they need them. Reassuring interior personnel that there's a way out and down is a positive move in helping them maintain focus. The officer or person in charge of the RIT should be performing their own size-up of the fire situation to determine progress and possible rescue needs in the event of a Mayday call. The RIT team should be doing a frequent walk around of the structure to see all sides of the building to determine where the rescue points are, what the progress of the incident is, and how it will affect rescue. Listening intently to the radio transmissions is another key indicator of interior progress and crew status. Elevated voices, rapid speaking, personal alert safety system (PASS) device alarms, and urgent messages are all indicators that the RIT team needs to be prepared for an imminent rescue, even though that may never happen. Some of these evolutions are policy, but others must be initiated by the company officer; however, that officer may be distracted or does not fully comprehend their duties and responsibilities as RIT. If the officer is distracted, others within the crew should speak up. Frequent training is also incumbent for success. The RIT team is there for us. It is our living, breathing insurance policy that we rely on as our firefighters in a pinch and should be treated as such. I have constantly drilled into my crews the RIT mindset: expect the unexpected!

FIG. 4–17. Multiple ground ladders were thrown (deployed) at a two-alarm tenement fire. (Photo courtesy of Jim Peruta)

Truck Placement

As with the mental mindset referenced in the St. Patrick's Day parade scenario, truck placement takes the same level of commitment to decision-making. Spotting the ladder apparatus is a skillset that is learned from mentors and experience. The apparatus operator must do what is necessary to affect rescue, maximize sweep and scrub area, and get the results needed to safety mitigate the situation.

These photos seen in figures 4–18a and b are of the same fire. One tower ladder is beached across the front lawn, the other backed down a tight driveway. Not easy tasks, especially with the 50' Sutphen tower, but the company officers and drivers of each knew what they had to do to affect the situation initially and they wasted no time delivering on that. When I arrived on the second alarm, I applauded both drivers and their officers for doing what is expected of them. They thought I was going to blast them for their decisions. Remember, your decision must be based on solid information and then you must retain the ability to justify it. Be ready and able to explain why you did something and what led to your decision to execute a given order. If you do something that is not spelled out in your department directives, you will probably be questioned on it, and may in fact be disciplined. However, firefighters understand that the dynamics of fireground operations are fluid, and sometimes what needs to be done cannot be found in a manual or guideline. I am in no way condoning rule breaking, freelancing, or reckless and irresponsible behavior. All I'm saying is that sometimes the human element needs to factor into decisions instead of algorithms. By the way, this type of truck placement is how I was trained to operate, and it has always been a definitive and acclaimed practice for Hartford Fire.

FIGS. 4–18a and b. Sometimes doing what may be viewed as unpopular is what is needed to affect rescue and mitigate the situation. (Photo courtesy of Pat Dooley)

The Minutia

Part 3 of the Connecticut Fire Safety Code section 1008.1.4.1 states revolving doors shall comply with the following directives: "#4 Each revolving door shall have a single hinged swinging door. . . in the same wall within 10 feet of the revolving door. A hotel occupancy (R1) requires a minimum of 2 doors."[11]

The point here is simple, yet elusive at times for some. Pay attention to details! If this little girl (my granddaughter) can build her Lego set compliant with Fire Safety and Building Codes (fig. 4–19), you as the officer or person in charge can certainly pay closer attention to the details of what you are doing. Glazing over specifics, omitting key points, being overwhelmed with broad strokes while simultaneously underwhelmed with details, and just plain laziness can lead to unwanted outcomes. Whatever endeavor you are engaged in, pay close attention to the details; they matter! Whether you're in the firehouse, dealing with the public or the administration, or at a working fire or technical incident, details count.

FIG. 4–19. My granddaughter pays attention to details. Do you?

---

11. State of Connecticut Fire Safety Code Part III: New Construction, Section 1008.1.4.1: Revolving Doors, 2012 ed.

Are you a Manager or a Leader?

Are you the type of officer that provides guidance, leadership, and mentoring? Or do you barricade yourself in your office for the entire shift only to come out for calls, food, bathroom breaks, and when your relief shows up? Are you taking the time to explain and teach the probies assigned to you and your house about the intricacies and nuances of the job? Or are you quick to dismiss inquiries and curiosities, rushing back to the barn so you can flop into the chair or on the bed and completely ignoring teachable moments for your crew (including the more senior members)? Are you setting the tone for the crew and the firehouse to create a productive and inclusive work environment? Or are you phoning it in, dumping it all on the senior firefighters to do your job, and then whining when things are screwed up? Are you laying the groundwork for everyone to be able to perform not only their own jobs but yours as well, so that when you are off, they know what to do, how to do it, and what the expected outcome is? Or are you overly reliant on the senior firefighter, the other officer in the house, and the overtime person, and you simply do not care because you don't have to be in that day? Do you understand the process and details of why we do certain things in the firehouse and on the fireground to better train, manage, and lead? Or do you simply do what is necessary to keep everyone off your back, not to ask your bosses questions, and to never get involved with the nuts and bolts of the department, and then turn around and brag on social media or to any unknowing listener about how you are the smartest person on the job and everyone else is the problem? As the officer, do you embrace your daily tasks in the firehouse and complete them diligently and thoroughly, taking initiative to complete expected goals like hose changing, ladder washing, window cleaning, waxing the apparatus, cutting grass, clearing the snow, reports, and other administrative paperwork? Or do you employ every excuse in the book to your relief to explain why you simply did not do your job the day before and how much of a rush you are in to leave the firehouse, dumping your incomplete work on that officer?

Are You Sweating Right About Now?

Are you the kind of officer that aspires to learn the job; get formally educated; speak with confidence, authority, and from experience; be taken seriously; and provide direction and guidance? Or do you really believe that simply showing up for work, going to a few fires, having fire academy certificates hanging on the wall, and wearing fire department shirts when off duty makes for a good firefighter? When you speak of incidents to which you have been, do you talk about what you have learned, what was effective and what was not, how decisions impacted safety, and what could have been done differently to minimize risk, while also appreciating others' input and passing on that knowledge and insight? Or are you the braggart who amplifies, embellishes, adds to, and morphs stories into hero-worship tales, who speaks in the first-person of things that did not happen and that you were not a part of, and who calls out those who performed poorly or made mistakes only to make yourself look better (fig. 4–20)?

When a member of your crew comes to you with an issue, such as payroll, their uniform or gear, or other administrative problems that pop up in the firehouse, do you take it seriously and get to work on it with the same diligence and speed at which you would solve your own issues? Would you then report back to that crew member what you are doing and what will be done to fix this problem while also surveying your crew for anyone else having this same issue that you could collectively solve together? Or are you just barebones managing the firehouse only to eat, sleep, watch TV, play video games, be the social media butterfly, and then go work overtime and do it all over again, having a personal philosophy that members are a bunch of crybabies and believing that you are not there to babysit them? Did you join the fire service, take the oath seriously, work hard to learn and perform the duties, garner experience, set personal goals and achievements, grow as a person, be a valuable part of your organization, share your knowledge to advance the mission and your department, and hand down a definitive contribution during your tenure? Or did you take the job simply for the cool uniforms, the convenient schedule, the inherent

FIG. 4–20. Every opportunity to teach can be considered mentoring. (Photo courtesy of Pat Dooley)

power trips, and the status that comes with the image and identity of a firefighter, only to further reinforce this with your big mouth and swollen head? As an officer, do you engage the human element of the mission with humility, compassion, and empathy for what the victims and their families may be going through? Or are you arrogant, self-absorbed, pretentious, single-minded, preoccupied with your own selfishness, and truly believe it is all about you? As an officer, when a member of the public comes to you with a concern, question, or gripe, do you engage that person or group with enthusiasm, professionalism, institutional knowledge and leadership skills; listen to them, obtain their contact info, and make every attempt to resolve their problem or connect them with someone who will; and then personally follow up to ensure their issues are being resolved? Or are you flippant, evasive, uninterested, not willing to help, dismissive, and intellectually unavailable; do you divert them to someone else who will just send them chasing their tail; or do you just not care? As an officer, do you speak from firsthand involvement, tested perseverance, and certified training, as well as acknowledge and appreciate that there is no substitute for experience? Or did you just hear it from someone and repeat it like a parrot, learn it in some class, regurgitate it from an internet video, or merely read it in a book?

Which Side Are You On?

Through my years of harvesting experience at actual incidents and situations, I have encountered each of the aforementioned scenarios in one form or another and have garnered an appreciation and understanding of how to be effectual and goal-oriented in personnel leadership beyond formal training and education. I am also proficient in what not to do and how not to interact with people, extracted primarily from individuals who simply did not have the best of intentions, the greatest attitude, or the most refined of personalities. One can learn equally from the bad as well as the good.

As an experienced fire officer, I am keenly aware of the lack of focus on the right way to handle human interactive situations and incidents in the fire service. I have observed great fireground commanders who were deficient when working with people, in a sense that it was taken for granted how to communicate effectively, including my less-than-desirable interactions, which were often the result of my approach to the situation. Fire academy training mainly consists of how to employ standardized manual-based theories and methods. Although well intentioned, these concepts often fall short when applied to the human element and frequently do not always match the situation, or worse, fail to deliver the desired outcome (fig. 4–21).

FIG. 4–21. Command presence is paramount to success. (Photo courtesy of Pat Dooley)

Although I was competent in the practice of management, as a new officer, I was not equally prepared in the leadership of people. Because of this inadequacy, I was often compelled to learn on the fly—the hard way. Throughout my career, most issues requiring formal investigations and disciplinary actions that I was either involved with or had knowledge of were primarily personnel issues, policy infractions, and contractual violations as opposed to operational deficiencies. These instances predominantly involved a rift with someone else in the firehouse or an individual not behaving in a professional manner. Just like on the fireground, you must assess and stabilize the situation immediately. To do so effectively, it's essential to identify the root cause to devise and execute a corrective action plan that is appropriate and satisfactory for all those involved. It is these experiences where a fire officer makes their bones and earns their keep, forging themselves as a leader (not just on the fireground).

The Right Choices

The solution here is simplistic in theory, but complex in execution. Many seasoned mechanics have said that the most abused and least maintained part of a car is the transmission. You can draw a direct parallel to the fire service, where the human interaction dynamic is the most utilized but the least trained-on part of the job. The solution is to put forth a conscious effort, determine your communication and interaction skillset, and know how you aspire to be received and perceived. To some, taking the right path comes naturally. For most, it requires a conscious effort, yet for some, it is simply beyond the scope of their capacity. The decision to do so and the subsequent communication process requires a solid foundation of the following intrinsic behaviors:

Build Soft Skills

Spend time with your crews to understand and appreciate who they are as well as what they do (walk the walk). The most important thing an officer can do for their crews is to spend time with them (the Golden Rule). Preserve ethics and compassion. Be the beacon of professionalism; do not just say you are. Think of each situation as if you were on the other side of it and notice how you would expect to be treated. Practice coherent communication and critical thinking. Listen intently to others and speak with respect and dignity, no matter how irrational or illogical you think they may be. Deliver your message objectively with clarity and purpose. Explain and teach rather than lecture and confuse. Affirm your command presence. If you are in a leadership capacity, act accordingly! The calming, trusted, inspirational authoritarian figure will always get things done. Trustworthiness and confidence instilled in others when they need it most, in all situations, is paramount. Maintain an open mind. Believe it or not, you do not know everything; you can learn something new every day and from anyone, regardless of age, rank, or tenure. Everyone has something unique to offer. Always strive to be a student of the fire service. Lead by example. You are compelled by your duty to set a positive example for others to follow. Do not squander those opportunities. Provide guidance and direction, not ambiguity. Embrace time management and the ability to be flexible in your decisions and, and most importantly, ensure your steadfast dependability. Leadership cannot exist without integrity. I was inadvertently thrust into a situation in which I had to inform my assignment of firefighters of the LODD of one of our crewmembers immediately following the incident. There was no one else to do it—just me. I was compelled to rise to the occasion and deliver on all the virtues of leadership.

Select the Proper Action

Do the right thing every time, especially when no one is watching. Choose the course of action for which you would not be embarrassed if it was reported on the front page of the daily newspaper. Set the tone. Not everyone you encounter can think at the same level or speak at the same pace as you. Communicating takes patience and diligence. Demand accountability. Own your actions and those that serve under you and embrace your responsibilities. Project the proper demeanor. Although you may have to speak to individuals with a different style for each, ensure the message remains the same. You do not have to take yourself seriously to take the job seriously. Maintain a positive attitude, especially when being criticized, and be gracious because this can be a teachable event for you. Always maintain emotional maturity and intelligence: if you cannot control yourself, your input may be rendered baseless by others. Know the job inside and out. How can you effectively lead if you yourself don't know? Learn the details and processes to understand the purpose. One of the fire officer's duties is to facilitate success for subordinates to fulfill the mission. Develop people ready for anything. Prepare subordinates for any inevitability. Embrace initiative and engage it instead of avoiding it. Demonstrate definitive problem-solving skills and the ability to acquire resources. Refine your ability to perform under the gun. I once trained my engine crew to be prepared for a bailout, an infrequently encountered event on the fireground. But just a day later, that same crew needed to bail out of a large fire while I was off duty. They called me at home and thanked me!

Take a World View

Maintain the big picture concept of how decisions and actions affect things and people beyond your immediacy. Build value, trustworthiness, and validity for both you and your people.

These attributes formulate a variety of tools in your toolbox from which to draw from and fulfill your duties and responsibilities as an officer. This is the truest sense of mentorship and succession development—the ability to confer transformational leadership which, in turn, facilitates those that have learned from you the capacity to teach others. The ancient Chinese philosopher Lao Tzu stated, "Give a man a fish and you feed him for a day. Teach him how to fish and you feed him for a lifetime."[12] Aptly applied, this lasting method of mentoring to produce the ability to succeed and thrive, even in your absence, is the ultimate goal: the lasting impact of enlightenment. But what good is all of this if you consistently default back to bad habits and the path of least resistance like a slow leak in a tire? The objective of achieving and maintaining this status is complicated, but not out of reach; it just takes dedication, commitment, and perseverance. Leadership can be taught, but true learning and comprehension (wisdom) comes from experience. New firefighters are exposed to knowledge and understanding that is not garnered in manuals but instead through real-world incidents and actual events, thus producing, sometimes unwittingly, the fundamental principles of how to simultaneously lead and manage. The inherent human interaction dynamic of fire officer duties demands providing a clear set of expectations, behaviors, and desired outcomes with personnel. This provides a roadmap for effective leadership capable of

---

12. Lau Tzu, *Inspiring Quotes* (blog), Pioneerthinking.com, June 28, 2013, https://pioneerthinking.com/teach-a-man-to-fish/.

achieving results in the firehouse and on the fireground. Without practical education and guidance through mentorship and succession development, those in leadership capacities are often cast into situations where they are charged with making critical decisions without the essential knowledge and skillsets. This lack of insight is often blamed for misinterpreting situations, which can have a ripple effect within the company and the entire shift. Often, excuses regarding the circumstances or being ill-prepared are made to justify negative outcomes. I have often focused my attention on the issue of properly leading and interacting with personnel through providing a practical understanding and appreciation for comprehension learned through experience and engagement; you know it as the stuff they don't teach you in school. Most firefighters know how to put out a fire, but it's not the fancy tools or shiny trucks that get it done; it's the human factor. Great firefighters do not always equal great leaders, but like anything else we do, a balance of training, education, and experience are the keys to success. However, let me be clear: some individuals may never be a great teacher, communicator, or leader, a fact that must be accepted and, to some extent, expected. When these tools are appropriately applied using this enlightened mindset, fire officers have the capacity to critically evaluate their situation, select the appropriate leadership strategies and tactics, and accomplish the intended outcome. Always remember that systems and situations are managed, but people need to be led![13]

Prepared versus Ready: Fundamental Differences

Why do some fire departments have the industry's best modern equipment, progressive policies and leadership, and virtually unlimited funding, yet still woefully underperform and underserve their community, struggle to meet their defined standard of coverage, and simply manage to consistently stumble, while other departments have the bare bones minimum of resources and equipment but consistently excel at their objectives both in their day-to-day operations and at incidents? Each is equally prepared and ready to engage and mitigate the situation at hand and carry out their missions, right? Maybe not. The answers lie deep within the core of the fire service: the human factor. An examination

---

13. Leigh H. Shapiro, "Fire Officers: Are You a Manager or a Leader?" *Leadership* (blog), Fire Engineering, April 2, 2019, https://www.fireengineering.com/leadership/fire-officers-are-you-a-manager-or-a-leader/#gref.

of the fundamental differences of being prepared versus being ready within the realm of critical thinking reveals the following key points.

Prepared

Being prepared is simply that, a state of preparedness with the prerequisite resources and personnel. Most departments are prepared for all types of emergencies, situations and conditions based on their core mission, geographical location, and mandated standard of coverage. They are prepared with the following tools:

Training

Most midsize and larger departments have their own training divisions or training officers who manage training requirements, mandated certifications, and refresher courses. However, some smaller departments may not have the budget or personnel required to maintain their own training needs, thus requiring outsourcing. Many years ago, the fire service was viewed merely as a vocation or a job. Today's modern fire service has been rightfully elevated to a profession validated by science and academia. Training is one of the principal pillars, for without it our mission is severely compromised.

Certifications. For a department to maintain preparedness, certifications need to be obtained and often renewed such as emergency medical technician, Firefighter I and II, hazmat, and so on. Specialized training in specific disciplines like rope rescue, high angle, and confined space rescue also require yearly recertification.

Classes. Often there are specific training classes and instruction required for personnel such as how to utilize a specific piece of equipment or apparatus, or how to properly carry out a specific guideline or policy. Sexual harassment, diversity training, and community interaction are often presented to fire personnel to not only train to the proper conduct and expectations, but also to minimize liability for the jurisdiction.

Degrees. Some departments mandate college degrees or credits as part of their core requirements for certain promotions for officer-level or specific jobs like fire marshal, emergency management director, and so on, while other departments and agencies do not. Some may also require a degree or accumulated credit hours as a prequalification for hiring. Many fire officers have or are in the process of achieving a degree to bolster their pay, increase their proficiency, or attain a promotion, and are already or soon be in an executive capacity which requires postsecondary learning. Often the reason may be a combination of several of these factors as many simply strive for their own professional development. Years ago, a degree in the fire science field of study was looked upon as unnecessary and overkill because of the nature of the job. Today dozens of schools and institutions offer degree and certificate programs because the demand for education is more prevalent than ever before, based in part on the National Fire Academy's creation of the Fire and Emergency Services Higher Education curriculum and the professional development model.[14] Having a degree or multiple degrees elevates a firefighter to a higher level of knowledge and career validation and bolsters overall preparedness. It does not, however put the fire out!

Resources

Every fire service agency and all-hazards response department has in place their respective resources based on their stated mission and standard of coverage.

Equipment. Power and hand tools, breathing apparatus, water application devices such as hose, nozzles, and fittings, along with the multitude of other specific pieces of equipment are what a department needs to sustain their operations to fulfill their duties.

Apparatus. Apparatus and vehicles are necessary (obviously) to transport personnel, tools and equipment, and other types of gear for functionality at any given scene. Water pumps, aerial devices, rescue boats, all-terrain vehicles, specialty vehicles, and the like are all part of the complement of apparatus required for preparedness.

Support. Having all this equipment and apparatus requires some type of support system in place to operate and maintain each. Tools and equipment often break down or need updating and upgrading, motorized apparatus require frequent maintenance, and specialized equipment such as ladders and breathing apparatus require yearly testing and maintenance. Furthermore, there are specific perishables which need replenishment: self-contained breathing apparatus bottles which need refilling; fuel and oil for apparatus, vehicles, and power tools; EMS supplies; and if departments operate their own rehab vehicle, supplies like food, water, energy drinks, and the like need replacing when consumed. The support divisions within a fire department also provide an important

---

14. National Fire Academy, Fire and Emergency Services Higher Education and the National Professional Development Model, 2023, https://www.usfa.fema.gov/nfa/about/feshe/course-outlines.html.

element in preparedness. The administration oversees all operations and policies as well as the budgetary requirements to operate the agency.

Knowledge

The greatest asset any department has is its people! And with people, you have inherent knowledge learned from a variety of means, sources, and experiences.

Policies. A department is not worth much if there are no definitive policies, procedures, and guidelines in place to steer the ship, set the tone, and develop and maintain the proper environment and standards. My department refers to these as *department directives*. Whatever their name, they provide a roadmap of how to proceed. If it's not written down somewhere, problems will frequently manifest within the department with leadership and management. If the department is unionized and has a contract, these guidelines and directives are accepted as an extension of the CBA. A violation of a directive equates to a violation of the contract and may indeed initiate the progressive discipline process for the violator(s) and the grievance process for the administration.

Experience. A department rich in experience based on longevity, history, and tradition has a deep pool to draw from in terms of how a department is shaped, what the core values are, and what defines it as an agency. Conversely, a young, less tenured department often draws from book knowledge and training frequently. Therefore, the senior member is identified with value and reverence because of the informal leadership position they may hold based on tenure and experience. This should not be confused with the old timer who has 1 year of experience 20 times, as they say!

Institutional Knowledge. Along with experience comes institutional knowledge. A department steeped in tradition generally has a specific way of doing things based on years of tried-and-true experience. This knowledge is passed down to other departments and personnel to facilitate a high degree of proficiency, confidence, effectiveness, and efficiency. The processes and procedures learned in a department that has had years of experience such as the FDNY, Hartford, Boston, and other storied departments quite often serves as a basis of knowledge for smaller, less busy, and younger (both in age of workforce and years in existence) departments. It can be said that in a busy department a firefighter would do more in 1 year than a slower department would do in 5. I experienced this having spent the first 10 years of my career in the busy north end of the city, then being transferred to the equally busy south end of the city for the next 10 years where the action trended and migrated based on demographics and socio-economic factors. This thankfully provided a solid foundation of experience to draw from throughout my career, especially when I was promoted to chief officer.

Motivation

It's nice to have all these resources and things, but what good are they if you don't have the properly motivated personnel to operate them?

Sworn. Most paid career departments mandate that their personnel be sworn in at the initiation of their careers. In other words, unlike garbage collectors and other city and town employees, police officers and firefighters must take an oath and swear not only to uphold the state and city or town charter, but to do their jobs to the best of their trained ability regardless of their personal feelings, convictions, and limitations. This is the truest definition of a professional.

Collective Bargaining Agreement. Many career, combination, and volunteer on-call departments abide by a CBA based on the local jurisdictions labor laws and the International Association of Fire Fighters framework for labor contracts.[15] This labor agreement often stipulates how the formulation and application of rules, regulations, boundaries and parameters of authority and responsibility for each department member are defined. Department administrations also hold managerial rights, in which other stipulations not defined by a CBA are carried out. Some departments do not have a CBA and their policies are often defined within the standard operating procedures. Most discipline frameworks are based on the CBA, as is the benefits package, hours of work, types of uniforms, when training can take place, and which certifications are required to maintain continued employment. This is mission critical for many departments, because without the contract, the structural employee framework must be derived from the human resources department of the jurisdiction.

Rules. Along with the CBA are the department or agency rules and regulations which set boundaries and guidelines for personnel behaviors. Often, the city or town has its own specific set of personnel rules and regulations that must be adhered to in conjunction with department rules. Individual fire departments expound on these rules to further explain and incorporate all personnel within the department.

---

15. International Association of Fire Fighters, "Model Contract Language Manual," 2019, https://www.iaff.org/wp-content/uploads/2019/06/18160_LICB_Model_Contract.pdf.

Morality. So, what makes somebody do something beyond rules, regulations, oaths, and contracts? Primarily, it's their individual morality and the human ability to do the right thing. This is probably the biggest motivator of them all, in that people in general are always striving to do the right thing, and the fire service is no different. Going above and beyond the call of duty to ensure a positive outcome is at the heart of the fire service. But don't be mistaken: people often falter, make poor decisions, and lack proper judgment, which is why there are rules, regulations, and disciplinary guidelines (fig. 4–22).

Ready

Does each firefighter of all ranks have the willpower and fortitude to act when called upon in an instant? Are they ready to work, know their job, and can do it? Do they follow the specific job performance requirements as spelled out in NFPA standards? Readiness cannot be taught or provided by the department: it comes from within an individual's core. Each of us must know our specific job thoroughly.

Willpower

It's more than just the duty to act; it's the drive to move forward and succeed at the tactical task at hand and the overall strategic mission. "Do you have what it takes?" is the frequent query among firefighters. The ability to engage, drive the mission, and see it through is the secret sauce in the recipe we call firefighting. True leadership derives from the depths of willpower, and without it, engagement of firefighting evolutions become ineffective, less than efficient, and often unnecessarily dangerous. Risk-taking is inherently built into our jobs; however, a comprehensive risk-versus-reward analysis should be first undertaken to influence decision-making. Years ago, I asked my chief of department

FIG. 4–22. A firefighter (my son Quentin) wearing a full protective ensemble, prepared to enter a structure and engage his duties. (Photo courtesy of Pat Dooley)

why we were not doing more real-world relevant training that would make us more proficient. Why are we not tapping into our internal resources (infrastructure and personnel) to move our mission forward and make training interesting and viable instead of something most of our personnel try, creatively at times, to avoid? He immediately blamed the dysfunction on the lack of proper funding. In frustration I quipped at his blanket excuse and responded, "If you had a stack of hundred-dollar bills from floor to ceiling you still couldn't get anything done!" It's not always about money; it's about the will to make it happen any way you can! He was not happy with me, but the truth can be difficult to ingest at times.

Fortitude to Act

It's one thing to know what and how to do it, but actually doing it is something completely different! Do you have the fortitude to act, meaning the mental and emotional strength and maturity in facing difficulty, adversity, and danger courageously (fig. 4–23)? Are you willing to make sacrifices to act according to expectations? Do you engage and act or do you cower and run? A senior tour commander once told me while he was preparing the daily personnel line-up on each apparatus that "We have firefighters, and we have the help!," meaning not everyone in the department was a superstar; some just worked there.

Ready to Work

It's nice of you to just show up, but are you ready to engage in your duties without a warmup session, mental preparation, or a pre-game group hug? When firefighters arrive on scene and the officers set the game plan into motion, they immediately engage in their duties and go to work. There's no pre-fire huddle

FIG. 4–23. My son Ashley taking a blow after a knockdown (Photo courtesy of Pat Dooley)

or handouts given out to crews; just go to work because you are ready to work! I once requested an additional fire company to one of my larger motor vehicle accident scenes: we had multiple victims in various states of need, a severe entrapment, a vehicle on fire and a significant hazmat condition. When the company arrived on scene, the captain walked up to the command post wearing no gear—just his work uniform—and asked, "What do you want us to do, Chief?" I snapped back at him to go put on his gear and then I'll tell him what I need. This created a delay in his company engaging in any duties at this scene. Later in the day I called him on the phone and told him forcefully "The next time you show up on one of my scenes, be ready to work—do you understand?" Message received and acknowledged. Just because he showed up didn't mean he was ready to perform.

Know the Job and Do It

To be in an effective state of readiness demands the officer knows the parameters of their authority, responsibility, and rank, which is paramount to success. To find the specifics, read your CBA thoroughly. Check your official job description (usually on the job announcement or in your contract under job descriptions). Become operationally proficient in the realm of building construction because that detail is often ignored or underserved. How can you fight the enemy if you don't know how it affects the structure you are in, or what if by engaging in suppression and overhaul operations you in fact destabilize and make unsafe an already fragile Jenga tower?

NFPA Job Performance Requirements

Within each NFPA standard are job performance requirements (JPRs).[16] JPRs take the mystery out of your duties and spell out in detail what is expected of you when you attain that specific level of training and certification. They explain what you should be able to do and how to go about doing it to successfully achieve this standard. For example, I responded to a large brush fire in one of our parks and found that my personnel were not familiar with how to put one of these types of fires out effectively other than throwing water on it. Later I reviewed the Firefighter I and II JPRs and proceeded to train them. It's written down somewhere!

What makes an effective and efficient team of firefighters? Is it the equipment and gear, policies and procedures, and all the other resources, or is there more to it? As clearly demonstrated, there is a huge difference between being prepared versus being ready. Suppose we are ready to engage in a task but find

---

16. NFPA. Job Performance Requirements (JPRs) and Knowledge, Skills, and Abilities (KSAs), nfpa.org.

we don't have the proper tools to do so; now what? Do we stop, reassess, then come up with a different strategy, or do we simply disengage and walk away? Throughout history, some of the most under-resourced and ill-prepared military operations were able to persevere because the soldiers and their leadership embodied the willpower and fortitude to accomplish their mission regardless of the obstacles they faced. Whether you can credit their leadership, or simply the spirit of the American soldier, one thing is clear: they get the job done! Similarly, the fire service has been the energy driving some of the most innovative and thoughtful inventions, creativity, foresight, and refinement of systems and processes worldwide. From breathing devices to extinguishment systems, from modern day tools and techniques to forward-thinking progressive policies, the fire service has consistently found itself compelled to adapt and improve for its own survival and the survival of others as well. This is what we do best, providing a foundation of dependability and consistency. When you call for help, we show up and make things better: no assembly required, batteries included, ready to work![17]

Case Study

When I was the senior tour commander on duty, an assignment was dispatched for a reported smoke condition in a high-rise during the day shift. Upon arrival, companies went to work looking for the source of alarm in conjunction with the caller who reported the smoke. The first-due engine and ladder companies found an elevator machinery room in the penthouse charged with smoke and heat from an overheated elevator motor and belt that caught fire. Since my chief's sports utility vehicle (SUV) was parked directedly across the street from the incident building, I remained in the front seat and used the radios and mobile data computer as a workstation and command post. Plus, because I was secluded from outside noise and interference at this crowded downtown incident, I could effortlessly speak on the radio and direct operations. While conducting my official duties running this incident, an older woman approached my passenger door to the SUV and began banging on the glass, asking if I could move so she could get her parked car out from her parking spot on the street. I had to ignore her because I was communicating on two different radio frequencies simultaneously, both with dispatch and with the on-scene companies. My aide, sitting next to me in the driver's seat was on the cellphone giving the chief of

---

17. Leigh H. Shapiro, "Prepared vs. Ready: Fundamental Differences for Firefighters," *Firefighter Nation* (blog), May 5, 2022, https://www.firefighternation.com/firerescue/prepared-vs-ready-fundamental-differences-for-firefighters/#gref.

department an update on the status of the situation (downtown incidents inherently bring extra attention). This woman banged on my window even harder and began to berate me for not getting out of her way so she could move her car. This incident was serious in nature, and I could not be distracted from communicating with forward companies operating in the fire room. The woman grew more impatient and belligerent until finally my aide jumped out of the car and went directly over to her and began to engage this woman. He was animated, and I could not hear what he said to her, but somehow this woman just walked away, and I did not see her again. Thankfully, the incident was mitigated, and I released the fire companies in a timely manner. My aide was notorious for his short fuse and larger-than-life temper, but always conducted himself professionally and with pride.

Questions

1. If you were the IC and this woman approached you with what you perceived as a ridiculous request, how would you proceed?
2. What steps would you employ to inform the public so that they fully understand exactly what we do and how we conduct our operations at an incident?
3. If that woman had a serious, legitimate reason why she needed to leave the scene in a hurry, how would you proceed?

5

The Incident Commander: Points to Enhance Skillsets

The incident commander (IC) is one of those functions that requires multiple proficiencies to be effective, including but not limited to exceptional organizational skills, emotional maturity, and the full knowledge and understanding of what each firefighter and crew is both engaged in and dealing with while operating on scene. A senior chief once quipped during a firehouse kitchen table discussion "You better know what you're doing because you may be ordering firefighters to their deaths!" Talk about the gravity of the job! A key skillset is possessing the respect of those you lead. Leadership and management must converge into an effective and efficient style capable of achieving not only strategic and tactical goals, but also communication while relaying confidence, authority, and responsibility. As the IC, I am expecting a result, not a process. I seek the end result of an incident based on safe practices and industry standards, comprehensive training, and definitive experience of those being lead and managed. I have learned early when standing in front of a blazing structure watching my plan unfold and form a cohesive intervention that I need to get comfortable being uncomfortable. The IC is not physically doing any work or evolutions. You are just standing there, sometimes animated, often agitated, barking directions to your crews. You are in the realm of the conceptual and the abstract, rather than the hands-on world you have become accustomed to as a firefighter and then company officer. For this text, I am referring to those chief officers and acting ICs whose function is to be the incident commander at a scene or scenario. A very wise young firefighter once told me that the most effective way to control an emergency incident as an IC is to turn the battlefield into a playing field. See what is happening and what you are doing conceptually as the boss in charge of all the moving parts and pieces, and embrace that which you can control, respect that which you cannot, and always be ready to pivot into any inevitability.

I see the IC as the conductor of an orchestra, waving the delicate baton at the troops while standing at the command post. The conductor themself does not actually make the music by playing an instrument. They just set the tone and tempo, guiding the musicians to follow the prearranged sheet music and turn it into something beautiful. The IC basically does the same. Fire crews already know what to do and how to achieve it; the IC is merely overseeing, directing, and managing those crews, and is not actually physically engaged in suppression evolutions. One could even argue that both the conductor of an orchestra and the IC of a fire are both doing a whole lot of nothing, but that is simply not the case. Both are intrinsically needed to function in a specific capacity for a definitive goal (fig. 5–1). I often refer to the IC as following the meatball recipe. When making meatballs, you obviously need specific ingredients such as meat, eggs, breadcrumbs, and whatever seasoning or flavor specific items you wish to include. Mix it together, separate into small portions and roll them into round shapes, like a ball. The fire service does the same thing across the spectrum of departments nationwide. They each have equipment and apparatus, firefighters and officers, and the mandate to save lives and property. How each department achieves this goal varies from one agency to another, but at the end of the day, they all save lives and property. Each meatball recipe varies from chef to chef, but at the end of the day, it's still a meatball, ready to eat. The fire service has the same mandate across all jurisdictions, we just do it differently in each one. Some do it one way, others another, but it gets done.

FIG. 5–1. The chief officer or IC's dress cap and helmet at the ready. Whether on the incident operations ground or interacting with the public, the projection of authority and command presence is always paramount to success.

The Street Boss

I have to say, I was extremely fortunate to have the set of circumstances in my career when I was promoted to captain. In Hartford, after serving your 6-month probationary period immediately following your promotion to captain, you are eligible to be detailed to the acting position of district chief (fig. 5–2).[1] You act and function in the same capacity as a ranking chief officer, responding to calls in the chief's sports utility vehicle (SUV), as well as managing and leading the crews in the city's north end. When the south end district chief is busy at a call, you are dispatched to cover his calls as well. Immediately following my 6-month probation, I was detailed right away to the acting position of district chief in the north end, known as District 2. I would serve under the guidance of the senior tour commander, the deputy chief in District 1. Any issues I had which needed a higher-ranking officer were directed to District 1, as well as covering his calls when he was tied up. Generally, a detail to District 2, which was affectionately known as the Red Car, were for only a single tour; however, with attrition from retirements and the city slow walking the promotional process, I was fortunate to often be detailed for a month at a time, as well as when needed intermittently. Furthermore, a third chief officer position on the line, known as District 3 would periodically be placed into service for varying reasons such as a large storm, city coverage during a mutual aid situation, and whenever circumstances dictated, per the collective bargaining agreement[2] and the chief of department. Since being promoted to captain in August of 2000, I had the fortunate opportunity to spend a lot of time serving in the Red Car as the acting district chief. It was 10 years of cutting my teeth, learning the job of chief officer, and burning things down, sometimes in spectacular fashion. I was so frequently detailed to the Red Car that I first earned the nickname Captain Chaos because I would frequently catch a lot of multiple-alarm working fires, extrications, and other types of calls requiring crews to go to work. I would look forward to being detailed as the district chief because as time went on, I felt more comfortable in that role, having done it so often, and because I was surrounded by senior personnel who kept a watchful eye on me and helped me to not only succeed, but to flourish in my duties.

---

1. "[Collective Bargaining] Agreement Between the City of Hartford and the Hartford Firefighters Association, July 1, 2009, through June 30, 2016," https://www.hartfordct.gov/files/assets/public/human-resources/hr-documents/hartford-fire-fighter-association-local-760-contract-7.1.2009-6.30.2016.pdf.
2. "[Collective Bargaining] Agreement Between the City of Hartford and the Hartford Firefighters Association, July 1, 2009, through June 30, 2016."

FIG. 5–2. A captain is eligible to be detailed to the acting job of district chief in Hartford. (Photo courtesy of Jim Peruta)

In 2010, I tested for the rank of deputy chief, and after scoring the highest and ranking number one on the list, I was promoted to deputy chief in March of 2011 and subsequently assigned to District 2 as the junior chief officer on shift. By then, I had acquired some serious trigger time as it's called, and although the rank was a new responsibility to me, the duties remained unchanged. I spent my first year as a deputy chief serving in the junior chief officer capacity of District 2. Then around April of the following year, I was transferred to District 1, which is the ranking deputy chief as the senior tour commander. Now I would be totally responsible for running the entire shift, managing the day-to-day operations, staffing, overtime, payroll, training, communicating with both internal and external resources and personnel, and interacting with the administration, as well as being dispatched to my own calls for service in the south end, covering the north end, and basically having my finger on the pulse of the city during my 24-hour tour.[3] It was a 24-hour job, with very little down time. My office phone would sometimes ring just as much at 4:00 a.m. as it would at 4:00 p.m. That's a city fire department for you!

Being the IC at an incident, specifically a working fire, is both a thrill and a challenge. There are a lot of moving parts that go into being an effective commander, both on the technical side and the human dynamic side as well. I learned a lot of tips and tricks to make life easier and have developed some methods that I will share with you here. From my years of experience as an

---

3. Hartford Fire Department, *District 1—Office of the Tour Commander Administrative Manual*, 2014.

IC, I have affectionately been called the "Street Boss" by my colleagues and peers, meaning that type of leader who is always both in the street and from the street, and is the actual boss of bosses, as opposed to a book chief, whose only experience is sitting at a desk or reading about someone else's experiences.

Let's get right into incident management, because this is where your knowledge, skills, and abilities need to be applied effectively to run an incident and get the results you are expecting. Hope is not a tactic! You cannot hope things will go well, or hope nobody gets hurt, or hope for anything. You must make it happen, sometimes by luck and preparedness, and sometimes by design or engineering. Either way, hope is not a factor in any of this. The IC, by design, is not only looking to mitigate the incident presented in front of them but should also be focused on several other key factors during the incident. Are the decisions made by the IC and company officers meeting the overall goal of effectiveness? Is what you're doing, or what you have the companies and crews doing, effective? Or is it what I have often heard from senior firefighters, a TW (timewaster)? Are you properly applying your resources to maximize your position and achieve your goals, or are you simply throwing things against the wall to see what may be? Is the overriding factor in all your decisions that of safety, or are you stuck between the rock of overly aggressive crews and the hard place of task paralysis because you are being overly cautious and are too timid to make the hard calls? It's a comfort level you need to hone. What is the overall progress of the incident? Which direction is it going in, and what decision will you be compelled to make: continue with your original plan or pull everyone out and go defensive? Is the fire getting bigger or smaller, and what is the condition of your personnel and of the structure you are operating in or around (fig. 5–3)? Can they maintain effectiveness or are they taking a beating and need relief?

Do you have situational awareness as the IC? Do you know what is happening on all sides of the building and all floors? Are you receiving proper radio reports with trusted intel or just winging it? Do you know the status of both engaged companies and staged companies and what is at your disposal in the event you need something else? Do you have peripheral awareness also? Do you know what's happening in another part of the city or jurisdiction that warrants your attention even though you are not there and already engaged in this incident? For example, if you have the department's only rescue company at your incident and something significant is dispatched to other companies, and it turns out that other call is a rescue or extrication, look at what you have the rescue company doing at your call. Are their tasks complete and you are in overhaul and salvage mode, and this rescue company is now just extra manpower? Cut them loose and allow that company to do what it's intended for. If

FIG. 5–3. The IC observing stream effectiveness and changing conditions (Photo courtesy of Jim Peruta)

you don't need to tie up the entire fleet of ladder companies at your incident just because they were dispatched there, cut one loose to cover the remainder of the city. You should be aware of what is happening behind you while you are standing at the command post (peripheral awareness) (fig. 5–4).

The rule of thumb for primary considerations while in command for me has always been these questions: what do I have, where is it going, what do I need, and what could go wrong? Keep these questions in your mind and answer them with the presented information, then make your decisions to intervene based on those answers. This is an ongoing process. It doesn't end once you have satisfactorily answered the questions. If you have active fire and that gets knocked down, your situation has changed, therefore your assessment needs updating. Many fatal incidents occur after the fire has been suppressed, so due diligence and a keen awareness of exactly what your situation status is during the overhaul phase is critical. Don't get complacent or lured into a false comfort level because of a cessation in energetic and often chaotic activities. Danger lurks everywhere in our job.

I learned early on as I sharpened my IC skills that it's much better overall to get out in front of a situation early, rather than relinquishing your initiative and having to play catch-up and altering your strategy later. Be proactive early in an incident. If you have some prior knowledge of how this situation may transpire, or some institutional knowledge about the structure or geographical area in which you are operating, don't be bashful in calling for additional resources. Get them on the road now to minimize turnout time and travel response. You may, in fact, need them sooner than you think.

5 • The Incident Commander: Points to Enhance Skillsets   115

FIG. 5–4. As the IC, do you know what's happening all around you (peripheral awareness)? (Photo courtesy of Pat Dooley)

Eugene Dumont, one of the chief's aides that I frequently worked with, often used the phrase, "Better to be looking at them than looking for them!" when referring to additional resources. This cannot be any truer, especially when things start to go sideways and you need help. One lesson I learned was from Arnold Goldstein, an old-school lieutenant from years back. Another old-timer I was working with one day was describing Goldstein's situation that, unfortunately by his own making, did not go well and he found himself in a jackpot at this incident. Lieutenant Goldstein was the officer in charge of a single engine company dispatched to a call for service which, back in those days, would only trigger a single-engine response. Upon his arrival, he confirmed whatever was happening and told the dispatcher he could handle the situation without any additional resources. Then, something went wrong, and he quickly realized he needed the full response assignment because the incident had escalated quickly. After doing what he could initially to mitigate this situation, he was resigned to waiting in the street for the additional companies to arrive and go to work. As he was waiting for several minutes in the street, the situation grew increasingly worse, which prompted this radio message from him to the dispatcher: "I don't hear any sirens, dispatch, are they coming?!" As comical as that sounds now, imagine how he felt, anxiously waiting for help to arrive. Therefore, if you think you may have to escalate the incident to maintain your initiative and to preserve your strategic and tactical advantage to achieve your expected results, then do it! You can always turn them around and release those resources, if in fact, you don't need them. Better to be looking at them than to be looking for them!

When issuing orders, whether in person or over the radio, maintain the tone of leadership and authority (fig. 5–5). If the troops detect even a hint of

FIG. 5–5. Bob Stella's prized handiwork. The gold shield carries the gravity and weight of the rank.

wavering, uncertainty, or inconsistency, they will sense that and begin to question not only your orders but also the tasks they are performing. We have all been there and taken direction from a boss when, as you proceed with the given direction or task, you either mumble to yourself or openly state aloud, "He doesn't know what he's doing!" As the IC, you at times may, in fact, be incorrect, need additional information, modify your orders or strategy, or simply admit you are wrong, but command presence is paramount for effective leadership and management. Issue your orders with tact and commitment, then get out of their way and let them do their job. No need for you to pile on to their troubles by physically standing in the crew's way at an incident or micromanaging them via the radio. You issued orders, let them work. If you see something needs changing, then change it, but allow the companies to perform the tasks and tactics you have laid out and trust them to complete. I've always been told that micromanagement is a mechanism for the incompetent. You should know your job inside and out, and now it's time to issue your orders like a street boss. When the opportunity presents itself, you should also communicate the why or rationale to your crews as well. It doesn't have to be at every incident and for every order, but a well-informed crew will work more effectively, efficiently, and harder if they understand why the task they are given needs to be completed. For example, my ladder company was issued an order at a large Class 4 heavy timber industrial complex that was totally engulfed in fire. The IC ordered us to force entry through a single bay garage door located in the front of the building. He didn't elaborate as to what we were supposed to do after forcing entry, nor did he state what would be on the other side. He simply ordered my crew to force the garage door. Once we did, we were met with a brick wall on three sides. We reported back to the IC at the command post that our task was complete. He asked how it went, to which the senior man blurted out in frustration, "That was a TW!" He asked what that was, and the senior man stated forcefully, "A timewaster!," meaning the task had no value in its result. If the IC had quickly stated that he was looking for another entry point into the building, or that if we found one, we could assist an engine company with setting up an additional attack line or master stream, then we as the crew would have been well informed and assisted in completing more than just the tactical task of forcible entry, but of the overall strategy as well. Like I said, it's not the job of the IC to spoon feed you a reason for every order that they issue, but in certain circumstances, it makes for a more effective crew if they understand why they are being tasked with something (fig. 5–6).

When I was in the process of studying for my deputy chief's promotional exam, the reading list which the eligible candidates were issued contained a list of books, documents, and policies where the exam questions would be drawn from. There were several International Fire Service Training Association (IFSTA) books listed, from chief officer to essentials, building construction, emergency medical services, hazmat, and so on. Included was the pump operator and aerial operator IFSTA manuals. Since I was studying for a deputy chief's promotion and not a driver's test, I found that perplexing as to why those lower-level functions would be included on a chief's test. Although the chief officer oversees many varying aspects of the fire service, the specific job I was testing for was line deputy, not a staff job, meaning I would be assigned to a district in our city and respond to calls for service. One day I approached our union president and asked why those books were on the reading list for chief officer. He stated that the joint health and safety committee consisting of both union management personnel and fire administration personnel met with the city human resources department to determine what the reading list would consist of, and this was the result. They concluded that this exam would be

FIG. 5–6. The IC or chief officer should always be approachable and ready for questions from those aspiring to learn the job. (Photo courtesy of Pat Dooley)

based on best industry practices, procedures, and standards at that time. He further stated that the reason those drivers' books are on the exam is because as the chief officer, you need to have a comprehensive knowledge base of what the drivers' tasks are and what they are doing at the scene, from pumping procedures to aerial operations. The more well-rounded your information knowledge base is, the better candidate and, ultimately if promoted, the better chief officer you will make. I thought about that for a minute, and it began to make perfect sense. I really needed an all-encompassing body of knowledge to be the most effective chief officer I could be. After a grueling testing period consisting of a written and oral exam, I scored highest on the promotional list.

Standing at the command post while all the action is playing out in front of you is in and of itself a challenge. Firefighters are used to getting their hands dirty and doing something, anything, on the fireground, except standing around. Often for me it became frustrating attempting to effectively communicate both in person and via the portable radio when the chaos was happening, and I needed to remain steadfast in my focus, emotional control, and body language. If I'm freaking out and losing my composure, that surely is translated to the working crews and everyone seeing and listening to me. The famous fire service illustrator Paul Combs created an image shown in figure 5–7 that perfectly

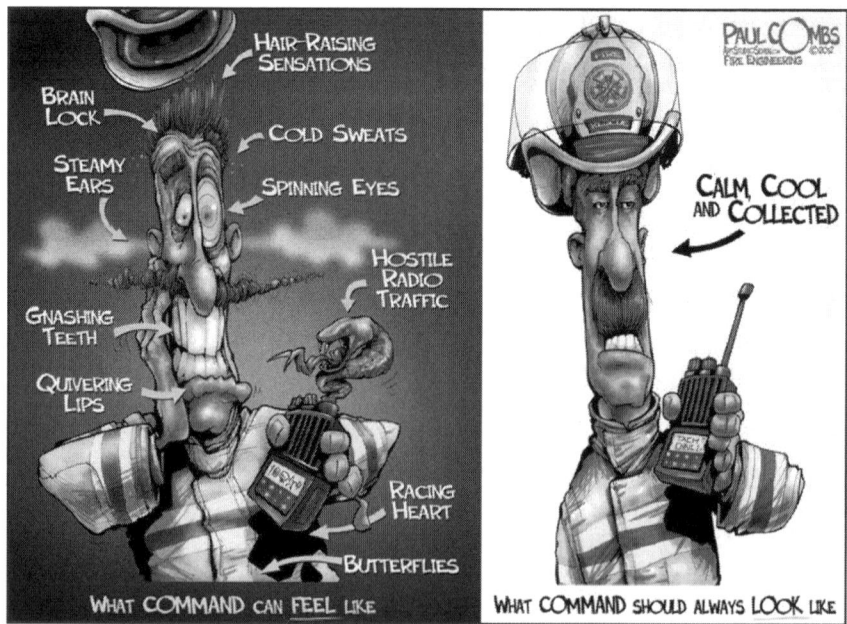

FIG. 5–7. Being in command looks and feels different depending on the point of view. (Illustration courtesy of Paul Combs)

captures in detail what incident command management should always look like, but what it can actually feel like as well, from the commander's standpoint.

Maintaining command presence on scene requires a calm, mentally collected, and cool-mannered individual. Speak clearly with a steady, unwavering tone and inflection, and project the image and identity of strong leadership, especially when things escalate and become strained. The troops depend on that. You are their rock, and the crews are the kite that is tied to that rock. You allow them the movement to perform, but you are steadfast in your post as the street boss. However, internally, the commander is fielding hostile radio traffic and compounding information that needs processing immediately. This often leads to physical feelings of a racing heart rate, quivering lips, gnashing teeth, steamy ears, cold sweats, hair-raising sensations, butterflies in your stomach, and an overwhelming sense of brain lock! Now obviously this is a humorous exaggeration of the signs and symptoms of stress; however, I can attest to having at least one or multiple symptoms during my tenure as a captain detailed to the acting chief's position, and especially as a chief officer. It's natural as a human being to react in this way, especially under tremendous stress and anxiety while operating at an incident. It doesn't even need to be a huge conflagration or multiple-alarm fire. It can be something simple yet extremely stressful, especially to the IC. Nevertheless, the duties of the street boss demand composure and leadership, regardless of what's happening. The truest definition of a professional is one who disregards their personal feelings and emotions to complete a task, duty, or obligation.

The Dance Card: Fireground Accountability for the Incident Commander, Simplified!

As I stated earlier, I had the fortunate opportunity to cut my teeth as the acting district chief for 10 years prior to being promoted to chief officer. That promotion solidified my duties as a full-time IC. I learned, often the hard way, what works, what doesn't, and who I could look to for solid advice, guidance, and above all, consistent mentorship (fig. 5–8). Once I became a deputy chief, serving as a district chief in charge of one end of, and eventually the entirety of, the city, it became clear that I had perfected my game to a comfort level I was able to easily apply with frequency and success. Please don't misunderstand what I am stating here. The method I will explain worked for me in my department based on how the resources, policies, and personnel were managed. It is

FIG. 5–8. Pictured standing with me is my deputy chief's aide, Anthony J. Guiliano, Jr., who, when I was assigned to the senior tour commanders' assignment, had been the senior chief's aide for several years already. It was the perfect synergy of time and grade with temperament to make for a more effective team. My job could never have been more effective and efficient if I did not have him by my side. A sage's wisdom, an encyclopedia's knowledge, and the heart of a lion is what defined my friend, Tony. Pictured in my hand is the famous dance card! (Photo courtesy of Pat Dooley)

only *a* way, not *the* way, and this worked best for me. You need to figure out what works best for you, what's permitted within your department, and your goals. It's also not very functional for large scale geographical incidents or complex situations with a lot of moving parts. When the IC is fixed at the command board by virtue of the nature of the incident, they are totally dependent on the information being fed to them. Therefore, radio reports must be concise and thorough instead of being full of chatter and devoid of any valid and actionable intel. The dance card was primarily designed for the common bread-and-butter-type incidents.

In the Hartford Fire Department, we are typically dispatched via the radio, which is transmitted both into the firehouse alert speaker system and over the mobile radio waves. It's one dispatch, not two like we had several years back. We first would get the dispatch in the firehouses via a Vocalarm system, which was a citywide internal intercom system, then over the radio. Eventually that antiquated system was removed, and the firehouse alert system was the same as the radio dispatch. When the trip lights were activated in the firehouse and you heard the dispatcher giving out the alarm information, it was simultaneously the radio transmission, but it could be heard in each firehouse that was being dispatched. Along with that, there were computers, both in the firehouse watch rooms and on each apparatus and chief's SUV, which would display the assignment information. The watch rooms were equipped with printers so the companies responding from quarters could grab the rip-and-run sheet to take with them. As the chief officer dispatched to multiple calls within my tour, the sheet became a valuable and integral asset for me. It shows on a single sheet of office paper the address, companies responding, and what is the reported problem, along with the specifics like time, date, box number, and National Fire Incident Reporting System number.

Pictured in figure 5–9 is a typical rip-and-run sheet. During our response, this sheet became a vital source of information for me because it allowed me to see what companies were responding to this call for service and which order they were responding as far as first due, second due, and so on. Often, the aide would take this sheet and transfer the information to the command board if the incident escalated and a command post needed to be established. If the sheet was not available, such as if we were not responding from quarters, a simple piece of folded paper with what companies and their response order using the exact same dispatch information on the computer screens in each apparatus and Red Car would suffice.

Figure 5–10 shows what a typical command board would look like at any of our incidents. Setting up the command board was job of the chief's aide, and it took several minutes to remove it from the chief's vehicle, bring it to the front of the building (which may be a distance away from the vehicle), physically set it up, and then begin to populate the incident information onto the board. For me, waiting for the command board to be set up and operational was a waste of valuable time. Besides, I was once told by my training chief that the command board was for everyone else at the incident. The IC needs to be engaged immediately. After working through this dilemma a few times, I came up with a solution that worked consistently for me, from the time the initial alarm sounded right up to when the command board was set up; from bell to board as I used to say! I call it the *dance card*!

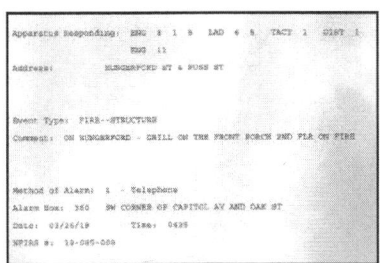

FIG. 5–9. A typical rip-and-run dispatch printout from the watchroom computer

FIG. 5–10. The command board at our incidents (Photo courtesy of Pat Dooley)

It's as simple as folding an 8.5" by 11" sheet of plain printer paper into four squares and copying what companies are responding and in what order on this assignment into a column. Look at the illustration of the rip-and-run sheet. It indicates that there is a reported fire on the second-floor front porch initiated by someone grilling. The building address is a common three-story brick ordinary legacy construction, typical in Hartford. I list my responding companies by their prearranged arrival order (fig. 5–11). The first-, second-, third-, and fourth-due engine companies are at the top, the rescue company in the middle (T stands for Tactical Unit 1 which is what Hartford calls their heavy rescue company), followed by the first- and second-due ladder companies at the bottom. This is the order in which our dispatches are sent out, per department directive. Then I simply populate this rectangular sheet of paper with the companies listed into the assignments I have given each company (fig. 5–12). The example shown indicates Engine 8 is assigned to the second floor as the first-in attack line, Engine 1 is second due and assigned to the floor above the fire with their handline, and Engine 5 is third due and assigned to the fire floor on number two to back up the first-due engine's handline. The fourth-due engine, per department directive, is the rapid intervention team, and the fifth-due engine is dispatched and assigned as the rehab unit upon confirmation of a working fire. The tactical unit, our heavy rescue company, is assigned to conduct the primary search, usually by splitting their four-person crew. The first-due ladder company is assigned to force entry and search, rescue, and vent on floor number two in conjunction with the two assigned engine companies, and the second-due ladder company is assigned to the roof for topside ventilation.[4] Before I even pull up to the incident address, I already know exactly where my companies are operating based on pre-arranged task assignments dictated by department directives, creating accountability and managing the incident effectively. Obviously, I would transfer command from the first-arriving company to myself to run the incident.

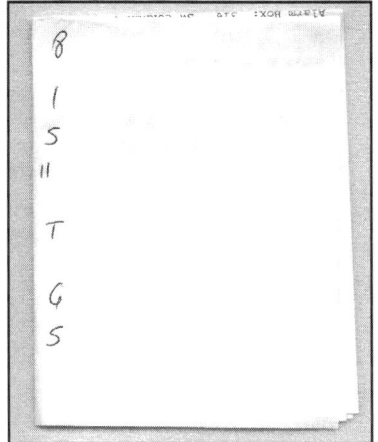

FIG. 5–11. The dance card: simplified fireground accountability from bell to board!

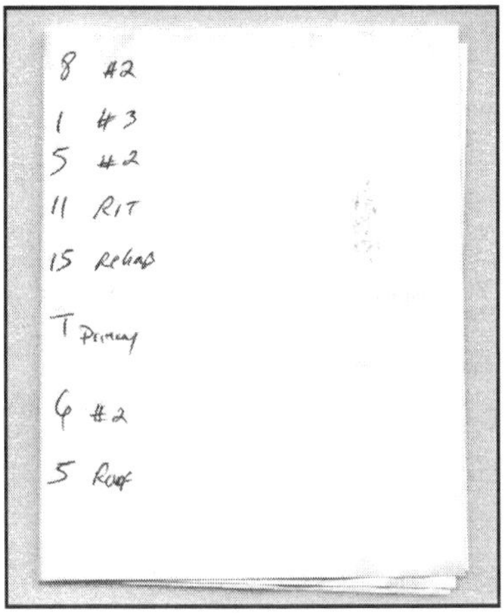

FIG. 5–12. The dance card with companies assigned (working)

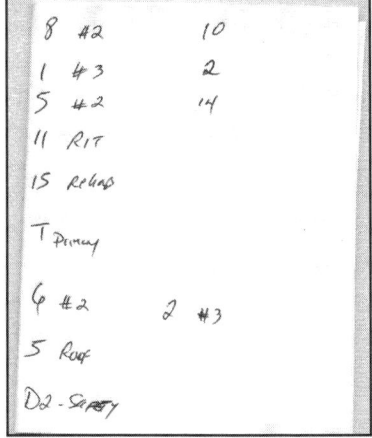

FIG. 5–13. The dance card with second alarm companies assigned

---

4. Hartford Fire Department, *Administrative Manual—Department Directives*, 2015.

With the dance card in hand, I can easily track the initial alarm assignment immediately, without waiting for any command board to be set up. It permits me to perform a continuous 360-degree size-up of the building to observe both conditions and progress. The command board can then be set up in a timely manner instead of putting everything on hold until it's established. The aide then can quickly transfer what I have transcribed onto the dance card onto the command board, allowing for uninterrupted command and control. If I need to escalate the incident by transmitting a second-alarm assignment, I simply add another column and assign those companies as well. By this example, the second-alarm assignment companies are listed as Engine 10 first due, Engine 2 second due, Engine 14 third due, and Ladder 2 first due, along with an additional chief officer of District 2 assigned as the safety officer per department directive, all on the second alarm (fig. 5–13).[5]

I have utilized this method in the IC role, and it is effective, efficient, expandible, and it just plain works, making an already complicated situation that much less complicated. It is simplistic in nature and has a proven track record. Over the years, several of the administrative chiefs have responded to my multiple-alarm incidents and asked me how I effectively managed the incident while not necessarily at the command board, even though my aide and the safety officer were stationed there. When I showed them, they were amazed that something so simple could be so effective and efficient. I like to use the analogy of the comparison between the Soviet Union space missions and the National Aeronautics and Space Administration's (NASA) space missions. NASA engineers spent millions of dollars and many years developing a pen that would write in a zero-gravity space environment during their missions. The Soviets simply used a pencil![6] Enough said.

As I stated earlier, the command board is for command, though I often viewed my role more of the operations chief when I ran an incident in that I was actively controlling the incident, partly to fulfill my strategic and tactical goals and partly to ensure incident safety. When we had major incidents, a dedicated command staff would be always positioned at the command post with the board. Although initiated at incidents, I would utilize the command board when I felt it was convenient. The board was dutifully populated and managed by my aide along with the second dispatched chief (safety) officer on scene. My operations accountability was already on the dance card, which is

---

5. Hartford Fire Department, *Administrative Manual—Department Directives*.
6. Ciara Curtin, "Fact or Fiction? NASA Spent Millions to Develop a Pen that Would Write in Space, whereas the Soviet Cosmonauts Used a Pencil," *Scientific American*, December 20, 2006, https://www.scientificamerican.com/article/fact-or-fiction-nasa-spen/.

always in my hand, therefore I could move about the fireground and observe real-time conditions firsthand, actively comparing what I was being told on the radio by the companies with what I was actually observing. This often made for a better, more fluid and effective command presence and helped facilitate whatever adjustments, modifications, and tweaks my incident action plan needed, including reinforcing working companies. Conditions changed and evolved since first arriving on scene when initial decisions were made, and I took full advantage of getting a personal view to blend incoming radio reports with personal observations of the fire building, operating personnel, smoke and fire conditions, intensity, and overall progress or lack of progress during suppression evolutions.[7] Remember, small or rural departments may find their command chief officer on an attack line instead of at a command post; your situation dictates what decisions must be made. Often, my rebuke of company reports based on what I was visualizing on the exterior made for a more effective company because it informed the officer of the effect their tasks may or may not have had on the progress of the operation. For example, I was communicating with an engine company officer during a working fire where that company was operating on the second floor with their line. Each time I asked the officer for a report, he kept insisting that their line was, in fact, hitting the fire and they were making headway. I told him via the radio that was not what I was seeing from the exterior and that they needed to either reposition or relocate their line. Turns out they were only hitting what they thought was the seat of the fire but, in fact, was only fire extension.

In Hartford, the district deputy chief is driven by a chief's aide, and that aide has many functions, one of which is to manage the command board during an incident (fig. 5–14).[8] By authority of the role they play, the aide often speaks with the commander's voice, meaning if they inform a company of a message or an order or they initiate a task, that ultimately emanates from the deputy chief. The military refers to this position as an *adjutant*. If the aide askes the dispatcher for resources or to perform a specific task, or if they inform a company to do something specific, that communication is an extension of the chief officer. Some company officers and support staff often chaffed at this; however, they usually were gently corrected and informed of the process. In this case, working with the senior chief's aide, he would not hesitate to inform me where I was lacking and deficient and would assist me in maintaining the correct track.

---

7. John J. Salka, Jr., "Strategy and Tactics," in *Chief Fire Officer's Desk Reference*, ed. John M. Buckman and the International Association of Fire Chiefs (Jones & Bartlett Learning, 2006).

8. Hartford Fire Department, *Administrative Manual—Department Directives*.

As figure 5–15 shows, my aide and I are always in constant communication with each other, even when he is managing the board and I am behind the building observing progress. We use the portable radios but more often face-to-face communication. He may leave the board and stand with me, having a second district chief and aide on scene manage the board. Where am I in figure 5–15? I'm watching the operations, not the board! To get technical, this

FIG. 5–14. Pictured is Senior Chief's Aide Anthony J. Giuliano, Jr. operating at the command board. Communications, board management, resource management and documentation are just some of the multiple duties and tasks the aide is responsible for.* Anthony was affectionately known as the Senior Red Car Driver of the United States of America! (Photo courtesy of Pat Dooley)

*Hartford Fire Department, *Administrative Manual—Department Directives.*

FIG. 5–15. I am observing operations and conditions with my aide, while other command staff officers are at the command board. (Photo courtesy of Jim Peruta)

method is aptly named *Management by Walking Around*, which can be found in its entire definition and explanation in any fire officer or business management textbook.[9] I am not condoning you abandoning the command board entirely. I am merely suggesting that when the opportunity presents itself, you should take advantage of it and get a good look around as to what is really happening at your incident. Anybody who is an experienced IC will testify that sometimes radio reports are lacking information, incorrect, or absent all together. The dance card method is just another technique to assist you in your duties. Besides, to me, standing in front of the command board is a lot like playing with the electric football board of years past (fig. 5–16).

FIG. 5–16. The command board in action! In this vintage electric football game, just when you set up all the players, the board vibrates and changes everything, much like the command board. The pieces are constantly moving. (Screenshot of video clip courtesy of Terence the DIYer)

The bottom line operating as the IC at a working fire is simple: you want to go from figure 5–17a to figure 5–17b! This fire at a large tenement was allowed to free burn for several minutes because the first-due engine officer did not follow procedure and bring up his high-rise standpipe equipment to the fire floor, as per department directive.[10] Instead, he stretched the limited hose supply from the apparatus' cross lay and obviously came up short, in effect, placing his company out of the game temporarily, thus allowing this fire to free burn for several minutes. A large volume of live fire converted to light colored active smoke and the reduction of turbulence and spread is a clear indicator that the line hitting the fire has had a positive affect and the situation is

---

9. IFSTA, *Chief Officer,* 2nd ed. (Fire Protection Publications, 2004), 176.

10. Hartford Fire Department, *Administrative Manual—Department Directives.*

FIGS. 5–17a and b. You want to go from this to this! (Photo courtesy of Jim Peruta)

going in the right direction. I often yelled out, "Touch down!" when I observed this condition change at a fire. It was a good feeling.

Simplified Format for Incident Management

I previously explained in depth a method for simplified fireground accountability; now I will explain another method I developed over the years to again simplify an already very complicated situation with many moving parts. It is a method for incident management whether it's a fire, vehicle crash, hazmat, or whatever you encounter. This format can be applied easily, allowing you to compartmentalize the order of the tasks and effectively manage the incident (fig. 5–18). I refer to it as the *size-up, set up, support* method.

Many times, when I have mentored junior officers on the skillsets needed for incident command, I often included formulating the mindset of thorough organizational skills. When I was a new captain having been detailed to the acting district chief position, I often found myself becoming overwhelmed at the monumental task of managing an incident. Even small incidents can have many challenges. "Where do I start?" is the frequent question from officers who are beginning to be exposed to running an incident larger than just their single engine or ladder company. The size-up, set up, support method addresses those key concerns and effectively provides an algorithm in the form of a

FIG. 5–18. Me, observing both building conditions and effectiveness of master steam application. (Photo courtesy of Pat Dooley)

template for the IC to follow, ensuring that your incident action plan (IAP) is thorough and minimizes the risk of missing something or getting confused with what needs to be done. This method breaks the incident down into three distinct stages, each one requiring specific actions by the IC, building from the first stage to the second stage and then the final stage. Each stage, however, demands constant reevaluation to protect the integrity of the IAP and meet both tactical and strategic goals while simultaneously allowing embedded flexibility to act and react to various dynamic variables.

Size-Up

An emergency medical technician would call size-up the *clinical impression*, which is a fancy term for, "What do you have in front of you?" and, "What's going on?" Based on the answer or partial answer to that question, the next question would be, "What do I need?" What do I need for resources to mitigate this situation? If I have enough resources to handle this situation, fine, but if not, now is the time to call for them. Remember, when speaking of resources, it is better to be looking at them than looking for them! The next question is, "Where is it going?" The "it" refers to the fire, hazmat, or whatever you are dealing with that could potentially grow, sometimes exponentially, causing you to lose your initiative. Your initiative is another word for your plan. And the king of all IC questions is, "What could go wrong?" It's all too easy to focus on the positive and set forth an IAP to mitigate the problem, but ask any seasoned IC and they will tell you that your plan B should include how to mitigate a situation that goes bad. What am I dealing with right now that could go wrong, throwing not only my plan into a tailspin, but require managing a whole new incident? For example, this could be a building collapse, a crew reporting they are lost or trapped, one building affected becoming two buildings affected, and so on and so on. What could go wrong? Having assessed this in your mind, you just need to know that there is the potential for something to happen and plan for it accordingly. You may not necessarily have to escalate the situation or the incident, but be prepared mentally in the event something goes wrong and try to base your current decisions on that mental state (fig. 5–19). An example is of an incident I was the IC for where we had a smoke condition deep inside a commercial structure requiring a crew to search for the source. The smoke was very dense, and visibility was near zero. I ordered the crew to come to the command post first before proceeding. In front of everyone at the board, I did a head count of who was going in and ordered a rope tag line placed on the officer. Once they began to proceed into the structure, I already knew several things: the radios probably won't work inside, it was every easy to get disoriented and lost, and retrieving this crew may be problematic. I already formulated my emergency action plan in the event something did go wrong, and all I had to do was pull the trigger and implement it. I addressed the "What could go wrong?" question by being proactive and ready, as opposed to reactive and unprepared. Having said all that, the incident priorities for a fire situation would be what I have utilized throughout my career: rescue, exposures, confinement, extinguish, overhaul, ventilation, and salvage (RECEO VS).[11] The order of these functions can be changed or even omitted depending on your

FIG. 5–19. The first stage is size-up; this is typically what you are confronted with when you roll up upon arrival. This also may not match with what you were told via the caller reporting, nor the first-arriving firefighter. (Photo courtesy of Pat Dooley)

---

11. IFSTA, *Chief Officer,* 400.

needs and the situation. Applying this template works well and can minimize confusion. Ventilation can be inserted wherever you deem appropriate.

Set Up

Simply put, set up is the "What are you doing?" phase (fig. 5–20). Now that you have hopefully identified what your situation is, you need to do something about it. In a hazmat incident, per the eight-step process,[12] you may in fact do nothing. The do-nothing plan is a possibility because of the nature of the hazmat, but let's stick with a building fire. So, what are you doing? What resources do you have with you, and do you need more? What is your strategy to mitigate this situation to a successful conclusion that ensures the safety of everyone involved? What tactics are you to employ to meet the goals of your strategy? What exactly are you doing now? This step involves deploying resources, initiating tasks, ordering crews, and placing all the puzzle pieces where they need to be to fulfill your plan, or "putting them to work," like I used to say.

Support

Now that you have executed whatever plan you came up with to fulfill your strategic and tactical goals, this is the phase where you keep it running (fig. 5–21). What are you going to do to maintain your plan and support the operation? Do you need additional resources above and beyond what was originally thought?

FIG. 5–20. The second stage is set up; by now you have a pretty good idea of what you are dealing with, so what is your plan? What do I do to set the resources and players into motion to initiate RECEO VS activities? (Photo courtesy of Pat Dooley)

---

12. Gregory Noll, Michael Hildebrand, and James Yvorra, *Hazardous Materials: Managing the Incident*, 3rd ed. (IFSTA, 2005).

FIG. 5–21. The third stage is support: What do I need to do to support the operation I just set into motion?; Do I need more suppression resources, or ancillary resources?; What notifications do I need to make?; and so on. (Photo courtesy of Jim Peruta)

Did things change requiring you to modify the plan? Do you need to call for any support services to assist with ensuring your IAP comes to fruition? For me, this was the time I would confirm that my aide was making the proper notifications, such as police, utility companies, our fire marshal's office, our special services division in charge of assisting those civilians affected by the fire, and whoever else needed to be notified for whatever reason. Sometimes it would be the health department, or a specific city department like Licenses and Inspections because we needed an electrical, mechanical, or building inspector on scene. This final phase ensures that the plan you set into motion continues to be supported.

Each one of these phases demands that you constantly reevaluate your situation and conditions and make the appropriate modifications to ensure safety and thoroughness. If you decide you need to escalate the situation, simply start over with the size-up, set up, support method. Now what do you have, now what are you doing, and now how will you support your operation?

The size-up, set-up, support method worked more efficiently for me than a standard incident command checksheet with boxes to fill-in and informational checkpoint elements scattered all over a page. Keep in mind, I'm running the incident like a street boss, not necessarily managing the incident like at the command board. Checksheets are good for individuals who like a bingo card style approach. For me, breaking the incident down into three sections worked best. Its all up to what you feel comfortable with.

The military adage of adapt and overcome is easily applied to what our mission is: to protect lives and property (fig. 5–22). If one thing in your IAP is allowed to halt the entire operation, then you will surely be ineffective at incident management. Even if during the incident the Mayday call is issued and the incident pivots to a rescue operation, the initial incident itself doesn't just go away, and you still have a fire to put out. In addressing less catastrophic situations, if for example a company relays to you that they have a dead hydrant for whatever reason (frozen, damaged, whatever), it does not mean that this company is out of the game. For the IC, it's about the results, not the excuses. Figure out a way to overcome, adapt, go around, or do whatever it takes to continue your plan and maintain your initiative. You can get creative at times, bend some rules, ask for thoughtful input, or even abandon a plan that simply cannot move forward without this now-defunct element in place. I'm buying the result, not the process.

FIG. 5–22. This dead hydrant in front of a major structure fire threw us a curve ball, but we adapted and overcame. We had three towers pounding this fire and water was at a premium: I directed an engine company to lay their supply line down the driveway and through the rear yards onto the next street over and hook up from a hydrant as close as possible. (Photo courtesy of Pat Dooley)

High-Rise Fires

For me, operating in a Class 1 high-rise building was relatively easy, considering what you're dealing with. That's not to say there won't be challenges, however. All I'm doing here is simplifying the obvious for you to gain the advantage. I equate high-rise operations to highway driving: everyone is going in the same direction and at relatively the same speed. Your playing field is all internal, so you must utilize each of the buildings built-in systems as tools to mitigate the incident.

I have had fires in each one of these structures (figs. 5–23a, b, c, and d), and many more high-rise buildings not shown here. Some fires were small, others more challenging; however, for all of them, I utilized the building systems to my advantage. Each one of these buildings has both active and passive fire protection systems, such as the ability to control the ventilation system, the ability to control the elevators, smoke dampers, sprinkler systems, standpipe systems, fire doors, pressurized stairwells, and whatever else may be in

Bushnell Towers—low building

777 Main Street (now residences)

SANDS high-rise

Betty Knox apartments (elderly)

FIGS. 5–23a, b, c, and d. High-rise buildings (Photos courtesy of Hartford Fire Department [HFD])

place.[13] All of these are tools for you to utilize to your advantage to gain the upper hand on a fire situation. These systems most likely are not available in an ordinary tenement or house fire, so you must act accordingly. Based on Hartford's operating procedures,[14] the first-due engine and ladder company reach the reported fire floor first, and their priority is to assess the problem, which is the size-up. The engine company is responsible for assessing fire conditions, and the ladder company is responsible for assessing ventilation and rescue conditions, all based on the "What do I have, what do I need, and where is it going?" model. The radio report back to the IC communicates this information so further strategic and tactical plans can be put into action. For the IC, this high-rise operation requires the size-up, set up, and support model to ensure thoroughness and completion of tasks. Much debate has been made about what nozzle the engine companies should be using during this firefight. Hartford Fire's directives mandate that a $^{15}/_{16}"$ smoothbore straight-tip nozzle be employed for adequate flow and penetration.[15] We learned early on in my department that you don't bring a knife to a gun fight!

Technical Rescue

Breaking a technical rescue down into manageable pieces is simple in that, besides following your department policies on how to manage and operate as the IC at this type of scenario, I found that knowing your specific role, and that of the other officers on scene, makes for a smooth, effective operation (figs. 5–24a and b). The IC is responsible for scene safety, the progress of the operation, managing and obtaining resources, and the result being a successful conclusion to the incident. The company officer, usually the rescue company, is the technical expert on scene, and I always allow them to do their thing unimpeded. I, as the IC, am there to support their operation and manage those elements listed previously. Quite often with these types of incidents, you get too many chefs in the kitchen so to speak, and it makes for an ineffective and cumbersome operation. My mantra? The rescue people are well trained and have lots of tools; let them work!

---

13. State of Connecticut Department of Administrative Services, "2022 Connecticut State Fire Safety Code," 2022, https://portal.ct.gov/-/media/DAS/Office-of-State-Building-Inspector/2022-State-Codes/2022-CSFSC-Final.pdf.
14. Hartford Fire Department, *Administrative Manual—Department Directives.*
15. Hartford Fire Department, *Administrative Manual—Department Directives.*

FIGS. 5–24a and b. The technical experts on the rescue unit have operational control of the situation, while the IC maintains command and safety. A rescue company member (my son) is rigging up a system in this photo. (Photo courtesy of HFD)

Case Study

The nature of our business is rudimentary, and therefore highly dependent on trust. The public trust in the fire service is paramount because when we show up to answer their desperate calls for help and they hand us an unresponsive infant or wave us into the house ablaze with all their worldly possessions inside, they are depending on us with blind trust that we are trained and ready to spring into action to save that child or that property. Volumes have been written and opined on being able to trust the IC, with all the common accompanying virtues bestowed on them. But what about the IC being able to trust those they command, communicate with, and ultimately depend on while operating at an incident? Trust. That's what it comes down to: the ability to trust others in their statements, their actions, and their results, and having confidence in their consistency and dependability. Without it, the IC can flounder, lose initiative, and be handed a result that was clearly not intended. The IC is very much like the conductor of an orchestra, trusting that each musician can effectively play their instrument to formulate what is required to produce music. Trust is obviously a two-way street, but as the IC, you must maintain the ability to exercise your trust in key people, crews, or companies on an assignment to get the job done. For the IC, the concept of trust is built and preserved based on several factors. When interacting with officers or persons in charge, what are your previous personal and professional interactions, if any? How has one or several individuals handled prior incidents and situations? How have they handled privileged information when the opportunity presented itself? Are the interactive radio reports you receive on scene valid and relevant? Beyond the usual academic leadership and trust bullet points, do you have a connection with the individual you are communicating with? Is there a disconnection and you are left to figure out for yourself what they may be telling or omitting during an incident? Are they minimizing or maximizing key information because that's how they perceive it or are they playing to a different audience? Do certain individuals or crews constantly need extra attention, coaching, and guidance to succeed, if at all? Have you built relationships with the people you interact with? Has anything you have taught or demonstrated to them had any influence on their conduct? Many firefighters can be habitual bunglers of information and communication and offer you what often seems like an alternate reality in terms of radio and face-to-face reports. Not everyone on the job is a superstar; some may only be there for a steady paycheck, and others may see it more as an occupation or vocation rather than a career or a calling, which is not to denigrate someone's motivation, only to illuminate a mental state of distraction and disengagement.

So, what does it take to construct that connection needed for implicit trust? How is it measured from one individual to another and how does it overcome the zero-credibility factor? The answers are not so simple. Although hope is certainly not a tactic in the IC's toolbox, there is, however, a lot of anticipation. One effective method I frequently utilized while serving as an IC at various incidents was to employ the method aptly named Management by Walking Around,[16] which can be found in its entire definition in any fire officer or business management textbook. Much like when President Ronald Reagan once remarked while negotiating with the Soviet Union, "Trust but verify,"[17] I would walk around the fireground scene and compare what I was being told via the radio or face to face with what I was seeing and make my own determination if in fact the information was accurate and actionable. Often the crews performing suppression activities are only privy to what they are directly engaging in, and do not see other evolutions playing out as well as changing conditions.

It's not that I distrust my officers and firefighters; this is merely a simple a way for me to get it right rather than being right and it ensures that I can receive the most accurate information to bring the situation to a safe and effective conclusion. The IC is dependent on the individuals' training, education, and experience to guide them to a successful outcome, because, as is obvious, once they (the fire crews) enter the building, my eyes are off them, and my entire interaction now may primarily become an operation of abstract images and concepts based on my own training, education, and experience. I must now believe they are doing the right thing and cautiously embrace my own level of intuitiveness. I must trust. Being the IC is not solely about formulating strategies and directing companies; effective communication is paramount. As I have often stated, your reputational value is a commodity and should be cultivated and utilized as such. Status and character may be frequently challenged, and having a less-than-stellar standing follows you around like a foul odor and can ultimately fuel distrust and prohibit you from certain privileges. Layers of confidence should be deliberately established, as often depicted in the phrase *win my trust*. Each instance builds onto another until there is a defined body of trust.

Ultimately, you are compelled to figure out how to interact and work with everyone, at all levels, to accomplish your goals and objectives, because it will always be about the result, not the process. Many times, the decisions made by the IC are both strategic and necessary to complete the task, the operation, or the entire incident. Some may argue that putting the right people in the right

---

16. IFSTA, *Chief Officer*, 176.

17. Ronald Reagan, 40th President of the United States, remarks after signing the Intermediate-Range Nuclear Forces (INF) Treaty with Soviet Union President Mikhail Gorbachev, December 1987, https://www.youtube.com/watch?v=qwh2w7osIp4.

places is the best method; however, we don't often have that opportunity or those people. Trusting in those you are ordering on the fireground to understand direction and complete the tasks safely and effectively is always the defining goal. An example comes to mind to illustrate exactly what I am referring to: Recently, a weather event in my neighborhood caused a large tree in my yard to fall onto my neighbor's garage. It required immediate removal, so instead of calling my usual tree service I have used in the past (and trusted based on experience) and possibly wait a week, I opted to hire another contractor who agreed to remove it immediately. After watching the crew work for several hours, I concluded that I had made a terrible mistake in hiring them because of the way they operated. Yes, the tree was successfully removed in a timely manner, and they certainly cleaned up after themselves, but the sheer lack of safety precautions and total disregard for safe planning and execution of removal techniques was startling. Each time they engaged in cutting and removing the large tree parts, they were setting themselves up for disaster and extreme danger to their own safety, not to mention the increased potential destruction to the objects in the yard. Once completed, I was grateful nobody got hurt or killed, or more damage was created. It was that bad! If this was a fire company or several fire crews, I would have stopped their suppression activities and had them swapped out for other crews. This type of blatant disregard for safety is unacceptable in the fire service, which often leads to distrust of those individuals or crews who operate in this capacity and reinforces my expectation bias. But when the officers and firefighters adhere to an attitude of safety, perform their evolutions in an effective and efficient manner, and do not normalize the deviation from safety standards and procedures, my comfort level of trust is safeguarded based on their predictable results. However, even established trust needs to be nurtured because it is wholly based on consistency and dependability.

The bottom line is elementary: I cannot teach you how to trust someone, nor can I compel you to implicitly trust anyone. I can, however, illuminate the path forward for the IC standing in front of the building as the rapid oxidation slowly dismantles what was generally considered a sound structure. Trust, like learning, is where you find it. It is inherent for humans to want to trust each other, and incumbent for the IC to exercise that quality to bring the incident to a safe, successful conclusion. Without that trust, we are a less effective team, a murky respite for the public, and a diminished champion of our mission.[18]

---

18. Leigh H. Shapiro, "The Incident Commander's Dilemma," *Leadership* (blog), Fire Engineering, October 9, 2023, https://www.fireengineering.com/leadership/the-incident-commanders-dilemma/#gref.

Questions

1. How would you build trust with your personnel beyond the usual academic bullet points (the mortar between the bricks)?
2. How would you foster growth with your personnel to achieve the desired communication level and status you believe to be most effective?
3. What steps would you employ in a situation where there is no connection with the personnel you are interacting with on scene, as may often be the case, yet you still have goals and objectives to achieve to mitigate the incident safely and effectively?

6

Fireground Behaviors

When I was in the fire investigator training program for Connecticut's fire marshal certification process, the material began making more sense to me in an unanticipated context. The primary text is the *National Fire Protection Association (NFPA) 921: Guide for Fire and Explosion Investigations* (fig. 6–1).[1] Reviewing chapter 11, "Fire-Related Human Behavior,"[2] the relevance of the material regarding firefighting struck me. The text is directed at the "actions and omissions of people associated with the incident scene,"[3] meaning "occupant characteristics and decision-making" (i.e., the civilians). This chapter resonated with me because it clearly relates to firefighters who, as human beings, behave just as civilians do (regardless of training, education, and experience) when faced with factors either within or beyond their control. "Actions and omissions" stood out because it offers a broad yet obscure explanation for certain individual and crew fireground behaviors. As I reflected on this often in my career as a company officer and then as a chief officer, how many times have I forcefully asked myself, crews, and individuals, "What are you doing?!," "Why did you do that?!," or, more importantly (post-incident), "Why did that happen?!" I was trying to understand beyond the official reports, the evidence, and the witness accounts why things ended badly for firefighters at certain incidents. The question was, "How can we explain the unexplainable in firefighter behaviors?"

The NFPA 921 guidebook is the nationally accepted, best-practice, industry standard by which to conduct fire investigations.[4] Chapter 11, "Fire-Related Human Behavior," was derived from the 1994 U.S. Fire Administration publication of the same title, which is based on research conducted by specialists

---

1. *NFPA 921: Guide for Fire and Explosion Investigations* (Quincy, MA: NFPA, 2021).
2. NFPA 921.
3. NFPA 921.
4. NFPA 921.

in the fire scene analysis and human behavior fields.[5] The findings provide insight as to how people react to fire emergencies both as individuals and in groups.[6] The numerous factors that affect an individual's or a group's human behavior before, during, and after a fire can be identified as characteristic of an individual, the population groups, the physical setting, and the fire itself. Furthermore, how these factors interact with one another to influence human behavior is explained.[7] To best describe the correlation between civilian actions and reactions during a fire situation and how and why things end badly for firefighters at incidents, you must first embrace the principle of cognitive comprehension limitations[8] and the factors that cause, support, and facilitate this condition. The psychology of human behaviors and, more importantly, failures under duress can explain how firefighters are reflex trained much like pilots: In essence, we are trained to react to specific algorithmic expectations regardless of the obvious obstacles and limited information, yet failures or deficiencies still occur for both. The Federal Aviation Administration (FAA) refers to this as the *startle response*, which states that control would have been maintained during an emergency if the pilot's knowledge and skills were fully

FIG. 6-1. NFPA 921: Guide for Fire and Explosion Investigations (2017 ed.)

---

5. NFPA 921.
6. NFPA 921.
7. NFPA 921.
8. NFPA 921.

developed to prepare for (any) emergency.⁹ The FAA determined that with adequate training pilots can "practice overcoming the natural human tendency toward denial and rationalization" and may mitigate the incident properly. So, how do you explain that, with all the training the fire service receives, injuries and deaths still occur beyond the obvious situations such as collapse, entrapments, medical issues, and so forth? Focusing solely on firefighters, I found that chapter 11 references numerous factors that influence fireground behaviors and may assist in explaining them. Although written to explain civilian actions in a fire situation, it clearly applies to firefighters, too. This material, however, is not intended to denigrate firefighters; it is simply formulated to help us better understand and explain certain behaviors.

Factors Affecting Behavior

The Individual (11.3.1)

Many characteristics influence human behavior in a fire situation. Physiological factors affect an individual's abilities to recognize and accurately assess the presented hazards and to react and respond appropriately.[10]

Physical Limitations (11.3.1.1)

This includes age, mobility, physical disabilities, intoxication, and other circumstances that restrict or limit an individual's ability to take appropriate action.[11] How many firefighters do we know who are still on the line and operating with prior compromising injuries, who are taking some type of medication that may affect their cognitive ability, or, by their own admission, are saying, "I'm too old for this!"?

Cognitive Comprehension Limitations (11.3.1.2)

This is the holy grail, as it largely explains why things happen.[12] This section includes many components that affect firefighters directly. These limitations have the greatest impact on firefighters and are more likely to explain behaviors and can account for delayed and even inappropriate responses to hazards and conditions. Often, multiple factors are involved, such as alcohol or

---

9. FAA Aviation Safety, "Startle Response," January 2022, FAA, https://www.faa.gov/sites/faa.gov/files/2022-01/Startle%20Response.pdf.

10. NFPA 921.

11. NFPA 921.

12. NFPA 921.

prescribed drug consumption prior to a fire (preignition), then carbon monoxide (smoke) inhalation post ignition.

Age. Age affects how firefighters perform. Mobility problems start to crop up, especially if there are previous injuries. My older brother was a lieutenant on the job back in the mid-1990s when he was involved in a major motorcycle crash that almost killed him. Thankfully, he recovered and returned to work, but with all the metal pins, rods, and screws inserted into his bones to put him back together again, he was never the same. His mobility became more of an issue during the strenuous physical demands of the job, too, especially in the winter and as he aged.

Mental Comprehension Level. Forcefully awakened from a deep sleep, thrust onto a careening fire apparatus in the dark of night, and then arriving on scene to commence a vigorous physical workout of setting up equipment and operations on the fireground is demanding enough. But how many times have we been confused or not fully aware of what was happening to us and around us? If we were computers, this would be the exact moment for a reboot!

Level of Rest. Studies have shown that even moderate sleep deprivation has a similar impact on personal performance and, in some cases, mimics that of being legally intoxicated from alcohol. After almost 20 hours without sleep, performance was gauged to be the equivalent of 0.05% blood alcohol content. Studies also indicate that working long or irregular hours, as with shift and night work, directly correlates to fatigue and an increase in accidents.[13] I once worked a 24-hour tour on our tactical unit (heavy rescue) one New Year's Eve, and when the tour ended at 8:00 a.m., we had 25 runs under our belt, including a couple of vehicle extrications and fires. I was so exhausted that I barely made the drive home. I was delirious from fatigue! I should have gotten a ride home, but I was not in the proper state of mind to process that decision, never mind making critical life-and-death fireground decisions. This happens every day for some firefighters—we believe it is just part of the job to which we must adapt accordingly.

---

13. A. M. Williamson and Anne-Marie Feyer, "Moderate Sleep Deprivation Produces Impairments in Cognitive and Motor Performance Equivalent to Legally Prescribed Levels of Alcohol Intoxication," *Occupational and Environmental Medicine* 57, no. 10 (October 2000), https://doi.org/10.1136/oem.57.10.649.

Alcohol Use. This is still prevalent within the fire service. We know this is the unspoken condition and that help is available, but drug and alcohol use is still a problem.

Developmental Disabilities. Applicants and candidates for recruit school are processed through the exam components that include cognitive and agility tests, and after they pass, they receive a physical exam. However, generally there is no mental exam to determine mental disorders or illnesses because it's not cost effective, variably subjective, and frankly, fire departments are seeking qualified candidates, and not in the business of diagnosing possible mental disorders. Many small, rural, and cash-strapped departments steer clear of this endeavor and rely on the cognitive exams and test scores from the fire academy or certification process.

Inhalation of Smoke or Gases. This factor can explain some fireground behavior, usually when things are already starting to go sideways. The composition of modern smoke is much more toxic (in the amount of chemicals and the concentrations) than in fires of the past. Although we use self-contained breathing apparatus (SCBA) often enough, some firefighters are still exposed to (and inhale) toxic gases and smoke.

Familiarity with Physical Settings (11.3.1.3)
Firefighters are keenly aware of their disadvantage at an incident of not knowing the physical setting of the operating environment; therefore, preplanning and training are critical to safety and success. If an individual had a fire in their own home, they would instinctively be a better judge of the fire's development and progression than if they were lodging in a hotel room. It is basic human nature to be intimately familiar with your own surroundings, which gives you the advantage in an emergency. If you factor in the physical and cognitive limitations, that advantage diminishes significantly. Chapter 11 cites the example of someone appearing to be lost or disoriented within their own home during a fire, sometimes found deceased.[14] Most often, firefighters have no prior knowledge of the physical setting in which they are operating. Subsequently, when things start to go bad, we are already at a disadvantage and must rely heavily on training and experience. When a lack of training, lack of experience, and cognitive and physical limitations overtake the situation, a near miss or, worse, a tragedy, can occur.

---

14. NFPA 921.

Groups (11.3.2.2, 11.3.2.3)

Chapter 11 states there is more than just certain characteristics that influence human behavior in a fire situation.[15] Interaction between individuals often tends to influence how people behave as well.

Group Size

People in a group are less likely to acknowledge and react appropriately to the sensory cues an emergency presents; the larger the group, the more likely this is. Research indicates that group members delay their responses until others in the group acknowledge the sensory cues and react, because responsibility for taking action is diffused among the group. Has this ever happened to you at a large incident where things are happening at a rapidly dynamic rate, and someone oversees the group (crew), yet no one takes charge (even with fire officers present)?

Group Structure

Groups with a formalized structure that have defined, recognized leaders and authority figures often react more quickly and in an orderly manner (which may not always be appropriate). Research also indicates that interaction between these group members often results in a sense of responsibility for the group, such as warning group members of danger as opposed to warning strangers. Group cohesiveness often results in a more unified and orderly response to danger, which also applies to fire crews. We have all worked with crews that were tight and with those that were anything but!

Group Permanence

Fire crews work well together most often when they are familiar with each other, as you do with your regular crew (fig. 6–2). If others are introduced to the crew—such as overtime or detailed personnel, new recruits, and members newly transferred into the group or shift—it affects performance. How long the individuals within the group have interacted with each other influences behavior. Research indicates that more established groups, such as families and sports teams (like fire crews), tend to be more formalized and structured and react differently than a new or transient group, in which each individual presents conflicting behavior based on their own unique reactions and responses.

Roles and Norms (11.3.2.4)

Chapter 11 describes behaviors affected by group norms and roles, such as gender, social class, education, and occupational composition (fire service

---

15. NFPA 921.

FIG. 6–2. On-scene resources report to the command post. (Photo courtesy of Jim Peruta)

scalar hierarchy).[16] Gender plays a big part in that women tend to be more clear-headed and report a fire immediately, whereas males tend to delay reporting because they often would rather initiate mitigation efforts such as suppression. Women and men process and act on information differently; the fire service is no exception.

Characteristics of Physical Setting (11.3.3)

The characteristics of the physical setting where the fire occurs directly impact its development and spread. Subsequently, they also affect fire-related human behavior. The building's height, alarm and suppression systems, and number and location of exits affect behavior.

Location of Exits

Research indicates that if the location of available exits is unknown to occupants or not adequately identified, they experience confusion and increased anxiety. This applies to firefighters, too. My department suffered a line-of-duty death (LODD) in 2014.[17] As I was standing out in front of the fire building, a

---

16. NFPA 921.

17. Hartford Fire Department, Firefighter Kevin Bell line-of-duty death while operating at 598 Blue Hills Ave., October 7, 2014.

firefighter plunged headfirst out of a second-story window and landed in the bushes below. It was later determined that he did not know the way out, became disoriented, and, nearly blacking out, exited the only way he could.

Number of Exits

The number of available exits also impacts behavior because if there are blocked, restricted, or unprotected exits, this may expose the occupants to the smoke, heat, and fire. The same applies to suppression crews, which is why it is critical to create a secondary means of egress—for example, placing ground ladders when crews are operating in a structure.

Height of Structure

Based on research, section 11.3.3.3[18] states some people believe they are less safe in taller buildings during a fire. Although movies such as *The Towering Inferno* do not help this dilemma, by design, modern high-rises are some of the safest structures in which a fire may occur. The intrinsically built-in active and passive safety systems such as alarm, sprinkler, and standpipe systems, in conjunction with smoke control, pressurized stairwells, and the rest of the Class I elements, make for an overall safer, more controlled environment.

Fire Alarm Systems

The constant repetition of false alarms and malfunctions plays a key role in occupants not responding appropriately to an activated alarm, and the fire service constantly battles complacency with crews not prepared to engage because of repeated false alarms. This happened in my department with the first-due crew responding to an alarm activation in a convalescent home one winter night in 2003.[19] The crew had been back and forth repeatedly over the previous several days; when the alarm came in again, they got off the apparatus wearing just their bunker gear. After walking through the building to the far end in the rear, they asked the charge nurse at the desk what was going on, to which she replied, "There's a fire back there!" Ultimately, 16 occupants perished in that fire back there! The fire was venting to the outside and gave the crew no indication there was anything happening until the fire room door became uncontrolled; then we had a problem. The crew, as it so often happens, became complacent and walked in unprepared. I, too, have fallen for this too many times to count; this crew just got caught.

---

18. NFPA 921.

19. Hartford, Connecticut, Greenwood Convalescent Home fire, February 26, 2003.

Fire-Suppression Systems

The presence of sprinkler and standpipe systems and even a Class II or III house line in the hallway of a structure may affect behavior. It increases the margin of safety for occupants; they may think they have more time to respond appropriately to the incident. Firefighters know what to expect when entering a protected versus an unprotected structure. In an unprotected structure, suppression evolutions may be hampered and therefore be more rigorous and even unsuccessful.

Factors in Delayed or Inappropriate Responses

Characteristics of the Fire (11.3.4)

Fire-related human behavior is directly related to both an individual's and a group's perception of the hazard and the threats that a fire or other incident presents.[20] The fire or incident characteristics shape these perceptions and, hence, affect behavior.

Presence of Flames (11.3.4.1)

Although most individuals have an uneducated and uninformed perception of the presented hazard, perception is amplified with the presence of fire, which confirms the situation is not a false alarm.[21] Because they don't understand fire dynamics and behavior, they may not interpret a small fire as an immediate threat, as happened at the Station Nightclub fire in Rhode Island in 2003. Likewise, firefighters may see a small amount of fire and not immediately grasp the fact that a large amount of hidden fire is literally eating away the building's core and rapidly compromising its structural integrity and their safety. Some may be overly focused on the moth-to-a-flame dilemma as opposed to the unfolding big picture scope.

Presence of Smoke (11.3.4.2)

Like fire, a smoke condition affects behavior because of an ignorance of fire dynamics.[22] Thick black smoke appears and is perceived as an immediate threat, whereas light gray smoke may be perceived as no threat. Although well trained, many firefighters may be inexperienced or untrained in how to read smoke and smoke conditions. Ignorance and inexperience may skew decision-making, often leading to a domino effect of bad decisions.

---

20. NFPA 921.

21. NFPA 921.

22. NFPA 921.

Effects of Toxic Gases and Oxygen Depletion (11.3.4.3)

During a fire situation, civilians often inhale the byproducts of combustion, including the toxic gases in smoke.[23] The depletion of oxygen below 15%, because of the progression of the fire, coupled with those gases, is a recipe for changes in human perception and behavior. Section 11.3.4.3 indicates these changes manifest in delayed or inappropriate responses to the incident, along with a severe decrease in strength, stamina, mental acuity, and perceptual ability. Although firefighters use their SCBA in an immediately dangerous to life or health (IDLH) atmosphere, how many times have we heard stories of old-school smoke-eaters who did not wear them or who used cheaters in addition to the more common issues of premature mask removal, masks being dislodged because of some type of interference, or even a malfunction? Once a firefighter inhales the same smoke and gases that a civilian would, albeit under different circumstances, the same physical effect holds true.

Human Factors Related To Fire Spread (11.7.1)

The actions or omissions of the people present before and during a fire can significantly affect fire spread, either accelerating or slowing it.[24] These actions include opening or closing doors and windows and rescue, to name a few. Captain Nick Papa of the New Britain, CT, Fire Department recently published *Coordinating Ventilation: Supporting Extinguishment and Survivability*[25] in which he emphasizes the critical importance of managing the openings within a building to support the firefighting operations and victims' survivability; uncontrolled doors and windows within a fire compartment can be detrimental to the outcome.

Recognition and Response (11.8)

In a fire, an individual's survivability depends on their ability to recognize and appropriately respond to the presented hazard.[26] Both civilians and firefighters must be able to perceive the danger, make an appropriate decision to act, and execute that action.

---

23. NFPA 921.
24. NFPA 921.
25. Nicholas Papa, *Coordinating Ventilation: Supporting Extinguishment and Survivability* (Fire Engineering Books and Videos, 2021).
26. NFPA 921.

Perception of Danger (Sensory Cues; 11.8.1)
People become aware of a fire through several sensory cues.[27] Factors such as being awake, asleep, or impaired affect perceiving these cues. Impairment can be physical, mental, or the result of drugs or alcohol, the inhalation of toxic gases, or low oxygen levels. Often, being jolted from a deep sleep, as in a firehouse setting, can set the stage for a disoriented firefighter being thrust into a complex environment in which they are not fully ready to safely engage.

Sight. The direct view of smoke or fire, visual alarms, and the flicker of an indirect flame.

Sound. The sounds of crackling, glass breaking, audible alarms, voices shouting, crying children, and dogs barking.

Feel. Rising radiating temperatures, failing structural members, and the wash of smoke over the body.

Smell. The unmistakable odor of smoke and to an experienced firefighter, what it specifically smells like.

Decision to Act (Response; 11.8.2)
An individual's degree of impairment plays a major factor in the decision-making process on how to respond to the severity of the perceived danger.[28] Firefighters are well trained, yet this thought process still plays out based on the individual in that we don't know for certain which, if any, impairment influences our decision-making process.

Escape Factors (11.8.4)
Firefighters often decide to ignore sensory perception of danger cues to complete their task at hand—such as rescue or other suppression activities.[29] Chapter 11 states the success or failure of escaping a fire depends on many factors including identifiable escape routes (e.g., ground or aerial ladders, secondary stairwells, and fire escapes or porches), the distance to the escape routes, fire conditions, blocked escape paths, familiarity with the structure, and occupant impairments. When you consider all the physical hurdles firefighters must face in making informed decisions, and then throw on top of that the real possibility of some type of identified impairment, this can become a recipe for disaster.

---

27. NFPA 921.
28. NFPA 921.
29. NFPA 921.

Heuristic Techniques

A heuristic is a mental shortcut that allows us to make decisions quickly without having all the relevant information. They are rules of thumb that allow us to make a decision that has a high probability of being correct without having to think everything through.[30] Although there are numerous variations of heuristics, Salt Lake City, UT, Clinical Law Professor Ian McCammon found six heuristic traps skiers fall into, causing recreational skiing avalanche accidents.[31] I overlaid these six points on firefighter behavior and found the following startling similarities:

- **Familiarity:** The belief that our behavior is correct because we have done it before. Our brains use shortcuts and experiences to guide decisions, giving the appearance that familiarity indicates safety when, in fact, it may not. Firefighters often employ this technique from the senior member perspective, but sometimes there can be a stark contrast between being clever or lucky and sound decisions based on actual knowledge.
- **Acceptance:** This is engagement in activities that we believe will get us noticed or accepted by those we like or respect or by those whom we want to like us and respect us. Firefighters may follow other firefighters or officers into an unsafe condition, perform unsafe acts, or make riskier decisions to be accepted, often because others have done it.
- **Consistency and commitment:** To maintain consistency, once the initial decision has been made, subsequent decisions are easier. Our brains use this shortcut because completely reevaluating a plan takes time and effort. Firefighters often face limited time to make rational decisions based on the information they have. This all-in mentality can quickly become dangerous groupthink!
- **Expert halo:** Groups tend to follow a leader even when that person may not have the best decision-making skills or adequate experience but often projects a halo of expertise. Such a leader can lead a group into dangerous situations.

---

30. "Heuristics," Conceptually.org, 2021, https://conceptually.org/concepts/heuristics.
31. Ian McCammon, "Heuristic Traps in Recreational Avalanche Accidents: Evidence and Implications," Montana State University, 2004, https://arc.lib.montana.edu/snow-science/objects/issw-2002-244-251.pdf.

- **Tracks and scarcity:** This is the tendency to value resources or opportunities in proportion to the chance you may lose them, especially to a competitor. How often do we hear one fire company trying to outdo another? Pride takes over and we are often hearing someone declare, "We only have one shot at this!"
- **Social facilitation:** This is the tendency to believe that a behavior is correct because others are also doing it as well. A group that is confident in their skills and abilities based on their experiences tends to make riskier decisions or to follow other groups because following is easier than leading. Unfortunately, fire companies and crews often engage in this dangerous behavior, normalizing the deviance from policy right up until something bad happens.

Reflex training is constantly battling human nature. When cognitive comprehension and the ability to act accordingly are compromised, bad things can happen to firefighters. If there is a lack of focus on processing information or on safely executing a task, hesitation, or a complete inability to recognize the gravity of the situation, failure is more likely. Frequently, firefighters will attempt to correct an already forgone situation based in part on the application of heuristic techniques. A National Institute for Occupational Safety and Health study of an LODD in 2010 indicated that when the firefighter was found (deceased) in the stairwell, he still had his SCBA facepiece in place, yet his air tank was completely empty.[32] It is unknown if in fact the firefighter was unconscious prior to running out of air; however, wouldn't human nature make the firefighter quickly remove the mask when he thought he was out of air and in trouble? How do you explain this scenario after all the research indicating human nature more often takes over for clear-headed thinking in situations of distress? During my recruit training on SCBA in 1988, we were taught that if you ran out of air or had a serious malfunction, rip your mask off your face and take your chances during self-evacuation rather than suffocate to death! There can be numerous rationalizations based on the research indicated here, but we can only speculate to explain the reasoning. What is a normal fight-or-flight instinct to a civilian is fervently resisted and deliberately ignored by firefighters because it is our nature to stand our ground, react and respond, and problem-solve to a positive result; our brains override our bodies (because it is our job). The foundation for all firefighters is their training, education, and

---

32. National Institute for Occupational Safety and Health, "A Career Lieutenant and a Career Firefighter Found Unresponsive at a Residential Structure Fire—Connecticut," Fire Fighter Investigation and Prevention Program F2010-18, June 1, 2011, https://www.cdc.gov/niosh/fire/reports/face201018.html.

experience in conjunction with steadfast preparation for any inevitability. When intellectual clarity is challenged beyond capacity or the integrity of psychological lucidity is somehow compromised, however, the default mental setting is to revert to our core human behaviors. Based on my research, when applying chapter 11's "Fire-Related Human Behavior," combined with the application of heuristic techniques to firefighters and fireground behavior, a direct correlation emerges, formulating the basis for a clearer understanding between this material and firefighter behaviors (beyond training, education, and experience). Dissecting the psychology of human behavior under duress provides some insight as to why bad things happen frequently that are devoid of rational explanation. If able to recognize the immediate perilous situation at hand, firefighters may often question their own decision-making processes, especially whether they should initiate a Mayday. Several ideas rushed through my mind when I faced this situation: "I should solve the problem (that's why I'm here)" and "I don't want any backlash or to have others think that I don't know what I'm doing or can't handle it." Instead of calling a Mayday when I thought I was in trouble, I was fighting with my own instincts to do my job and complete my tasks, which had now metastasized into a functional bad habit.

Firefighter survival fundamentally depends on steadfast proficiencies consistent with knowledge, skills, discipline, mental focus, situational awareness, and the effective ability to apply critical thinking. Our training is in constant battle with our own human nature and, in the end, when pushed to our limits with cognitive factors beyond our control, ultimately, our ingrained and instinctual human behavior wins the day.[33]

Overcoming Three Common Fireground Failures

Whether you are a company officer or an incident commander (IC), you will inevitably be faced with challenges that have the potential to stop you in your tracks. Having been around a few years myself, I have learned to respect three types of fireground failures that tend to occur frequently and have propelled my crews and I into a creative problem-solving mindset so we could complete our mission. Firefighters thrive on professional challenges; just look at the number

---

33. Leigh H. Shapiro, "Fire(fighter)-Related Human Behavior: Understanding Cognitive Limitations on the Fireground," *Leadership* (blog), Fire Engineering, February 1, 2022, https://www.fireengineering.com/leadership/firefighter-related-human-behavior-cognitive-limitations/#gref.

of training publications, seminars, and classes designed to address these issues. What information is offered here that you cannot learn elsewhere? Well, these three fireground failures are so commonplace that they are perceived subliminally and may not register on our radar screen until we recognize the pattern.

Failure #1: The Plan

World War I German military strategist Helmuth von Moltke aptly stated, "No battle plan survives contact with the enemy,"[34] and he was right! Most of the time, a derivative of a stated plan works just fine, but overall, plans do not typically go as they were intended. For this reason, use a plan as a guideline instead of as a strict game plan. When you arrive on the incident scene, just the fact that you need additional resources or alarms is an indication that the initial action plan needs revision. Although it is necessary to maintain standard operating guidelines, procedures, and directives, do not allow yourself and your objectives to get locked into a set plan that does not or cannot consider the unexpected, unintended, or unwanted jolts of reality. It is equally imperative to be dynamic in your thought process by constantly performing an ongoing size-up and contingency planning. Ensure that the inevitable question, "Now what do I do?" is answered with a plan B. Through consistent reevaluation of the situation, even after fire knockdown, the questions "What do I have?"; "Where is it going?"; "What do I need?"; and "What could go wrong?" are readily anticipated (fig. 6–3).

FIG. 6–3. Companies arrive to find the structure well involved, forcing an immediate incident action plan revision.

---

34. "Helmuth von Moltke the Elder," Wikiqoute.com, 2023, https://en.wikiquote.org/wiki/Helmuth_von_Moltke_the_Elder.

Failure #2: Technology

Technology, in all its glory, will inevitably fail, likely when you need it the most (fig. 6–4). As the fire service evolves, so does complacency and laziness; we'd all love to have an app for that. Unfortunately, our job still needs to be done. The following statements (usually stated in the heat of the moment) are all too familiar to firefighters: "The dispatch info is wrong"; "The radios are junk"; "The computers are slow"; "The thermals are old"; "The meters are off"; "The batteries are dead"; "The saw won't start"; "It worked earlier" (my personal favorite), and so on. How many times while responding to or operating on the fireground have you muttered these statements or heard them so eloquently applied as the reasons something didn't go as planned?

Overreliance on personal protective equipment and electronics (fancy gadgets and gizmos) has led to a reduction in critical thinking and a suppression of the core senses, leading to disastrous fireground situations that are completely preventable. Even cavemen figured out how to deal with fire; so, what's changed? Technology, whether it's thermal imaging cameras, encapsulating turnout gear, or computers, should be used for their intended purpose: to assist and accentuate, not replace! Be ready for that old-school firefighting of using your training, your experience, and your gut (intuition) to guide you, and let the technology be what it was intended to be: another tool in your toolbox.

FIG. 6–4. A citywide radio failure prompted this error message while en route to an incident.

Failure #3: Initiative

Instead of operating as you originally planned, your actions have now transitioned into playing catch-up at the incident because you lost your initiative and now must do something different (fig. 6–5). Why? Because what you thought was a definite is wrong; what you believed was going well is not; what you were initially told is inaccurate or incomplete; what you thought would not happen just

FIG. 6–5. Residents tied bedsheets together to create an escape route out the back of the building during an apartment fire.

happened; equipment and people you depended on were not reliable; tools and crews did not perform as expected; the fire gets bigger, not smaller; the structure may weaken to the point of compromise; needed resources may not be available; or, the ultimate kicker, someone calls a Mayday! This all ties back into what I stated earlier about anticipating these potential issues and having alternative plans in place to address them. The basic strategy remains largely the same, but the tactics must adapt (in real time) to what is happening. Eventually, you will get to where you want to be, but not expecting the unexpected is the surest way to bring everything to a screeching halt. More important, however, is recognizing and overcoming these failures to complete your mission effectively and efficiently.

The Point

You may not have heard of these three fireground failures placed in this context before or, more likely, you never consciously realized how these simple things can undermine an operation, but when you spend more time fixing the tools rather than fixing the car, as my father used to say, you begin to see the pattern, which becomes frustrating and unnerving. This is the part of the job of a company officer or an IC that seems to always happen in one form or another, but it is often forgotten about at the successful conclusion of an incident because the job got done and we mentally categorized these occurrences as incidental. By ignoring or failing to address the root cause, we are potentially setting ourselves up for future failures. This oversight can eventually lead to a more severe or even tragic outcome. Just because the presumptions of plan implementation, functional technology, and initiative priorities are set in motion but don't go as expected doesn't mean you're out of the game. You must evaluate the situation and adjust accordingly. The fireground is a highly dynamic environment; therefore, we must remain aware of our surroundings and be flexible. During the World War II invasion of Normandy, Brigadier General Theodore Roosevelt Jr. found that his division's landing crafts had drifted south of their objective, placing the first wave of his men a mile off course. As they approached the beach, he assessed the situation and chose to fight from their current location rather than to attempt to relocate to their intended position. He stated, "We'll start the war from right here!"[35] The ability to adapt, improvise, and overcome unanticipated adversity is critical to mission success, and it's what we do best![36]

Interior Fire Attack: Obsolete or Indispensable?

Intermingled within the vast online world of fire service opinions, many of which are given life by various trade blogs, is an idea predicting the not-so-distant future of firefighting devoid of any interior fire attack whatsoever: working fires will be mitigated from the outside of the structure much like the days of old, prior to the adoption of SCBA and science-based suppression techniques. The thinking is that interior suppression evolutions are simply too dangerous, financially unsustainable, and based on antiquated operating models steeped more in tradition and pride than actual modern concepts.[37] This concept does not address the need for all existing buildings, houses, and structures to be nonexistent in the near future for some strange reason, thus creating a false panacea of safety rendered by the do-not-enter method. I submit to you a direct challenge to the entire premise of no interior firefighting ops, period! It is true that the constantly evolving changes in modern building construction materials

---

35. "Theodore Roosevelt Jr.," Wikipedia.com, 2023, https://en.wikipedia.org/wiki/Theodore_Roosevelt_Jr.
36. Leigh H. Shapiro, "Recognizing and Overcoming Three Common Fireground Failures," *Fireground Management* (blog), Fire Engineering, July 1, 2018, https://www.fireengineering.com/leadership/fireground-management/recognizing-and-overcoming-three-common-fireground-failures/#gref.
37. Robert Avsec, "Why Interior Firefighting Will Be Obsolete by 2030," *Fire & EMS Leader Pro* (blog), December 17, 2018, https://www.fireemsleaderpro.org/2018/12/17/interior-firefighting-obsolete-2030/.

and design concepts have had unwanted and sometimes fatal results for firefighters. Petroleum-based furnishings and contents, cheap and often substandard construction materials, radical construction concepts that are designed to fail so they can be cheaply rebuilt, open floor plans, increased fuel loads, and the all-important rising rate of cancer due to the inclusion of carcinogens are all unwelcomed aspects of contemporary firefighting. The compounding trend of personnel affected by cancer and the associated death rates demands immediate attention at all levels of the fire service, especially operational models, methods, and policies. However, the fire service's mission is the foundation of what we do, how we do it and why, and the expected outcomes. Just as the military is constantly modernizing and increasingly using remote operated and unmanned platforms to minimize inherent risk to soldiers and optimize outcomes, so too has the fire service adopted similar changes like robotics, drones, and computer-based technology. One only has to look as far as the 2019 Notre Dame Cathedral fire in Paris, France, to see drones and robots deployed to assist in suppression evolutions, thus initially keeping suppression forces out of harm's way.[38] The key word here is assist, not replace. Purveyors of the exterior ops-only concept create a new problem in their limited and overly simplistic attempt to solve an old one: many fire departments short on resources or other operational hindrances may become overly reliant on the exterior extinguishment method. However, the primary element this hit-it-hard-from-the-yard fire suppression concept is missing is the human factor. Conceptually, the incident action plan forgoes primary and secondary searches for occupants, thorough overhaul activities, and careful salvage actions. Firefighters cannot write off every building fire no matter the size and scope because the tasks are too dangerous. The cost both in financial and humanitarian terms would be incalculable. The modern-day mindset of safety at all costs has become a beacon for this operational model of ineffectiveness and inefficiency devoid of any application of common sense and critical thinking. This overbearing, broad-brushed mentality of risk aversion is problematic at best, and dangerous at worst!

Let's assume this paradigm shift is adopted in a municipal professional fire department generally known and admired for their aggressive interior fire attack methods. How long do you think this new practice would be acceptable until it became unacceptable? Insurance companies covering property and casualty losses would be screaming about how the fire department turned a simple food-on-the-stove fire into a multi-million-dollar loss or created a billion-dollar catastrophic event due to the inaction of responding fire

---

38. Steve Crow, "How Drones and Robots Helped Extinguish Notre Dame Fire," *The Robot Report* (blog), April 18, 2019, https://www.therobotreport.com/how-drones-robots-helped-extinguish-notre-dame-fire/.

companies to initiate mitigating measures (which is why we're here in the first place). This strategy would bankrupt an already financially strapped municipality because it would be repeatedly sued for knowingly allowing rescuable people to die over and over due to its fire department's policy of exterior-only operations at working fires. There would be mobs with pitchforks and torches in front of city hall demanding the fire chief, the mayor, and anybody else connected to this ridiculous policy be fired immediately for allowing this to occur and failing to protect the residents, workers, visitors, and property within the municipality. A December 26, 2015, fire in San Luis Obispo, California, has entangled their fire department in litigation because of the adoption and application of an exterior operations-only method of structural firefighting (Case #Sub v SLO 18CV-0782).[39] The plaintiffs (the property owners and insurance company) contend the fire department, predicated on this strategy, "purposefully burned down the structure rather than put firefighters in harm's way!" While research at Underwriters Laboratories (UL) and the National Institute of Standards and Technology has shown transitional attack to be a viable option for rapid knockdown of a large volume of fire, the January 2018 UL Firefighter Safety Research Institute report titled *Impact of Fire Attack Utilizing Interior and Exterior Streams on Firefighter Safety and Occupant Survival: Full Scale Experiments* states the transitional knock-back method of exterior fire control with the proper nozzle selection, as tested in a controlled environment, is safe and effective for modern-day suppression evolutions.[40] In fact, the entire report debunks any myth of exterior-only fire attack. This transitional fire attack method is widely considered to be an acceptable practice today as the way of first attacking a large volume of fire to advance suppression evolutions to their conclusion of putting out the fire. The premise of exterior-only operations, however, is rooted in the political bean counter mentality of junk science having no basis in the real world of fire science. Applying the Daubert Standard method,[41] the no-interior-firefighting concept does not hold up under scrutiny:

---

39. Curt Varone, "San Luis Obispo Sued for $5 Million Over 2015 Fire," *Fire Law Blog* (blog), December 30, 2018, https://www.firelawblog.com/2018/12/30/san-luis-obispo-sued-for-5-million-over-2015-fire/.

40. Robin Zevotek, Keith Stakes, and Joseph Willi, "Impact of Fire Attack Utilizing Interior and Exterior Streams on Firefighter Safety and Occupant Survival: Full Scale Experiments," Underwriters Laboratories Fire Safety Research Institute, January 25, 2018, https://technicalpanels.fsri.org/docs/DHS2013_Part_III_Full_Scale.pdf.

41. "The Daubert Test," Public Health Law Map—Beta 5.7, April 19, 2009, https://biotech.law.lsu.edu/map/TheDaubertTest.html.

1. Has the new concept been extensively tested?
2. What are the rates of error when applied?
3. Has it been subjected to peer review?
4. What control standards is it based upon?
5. Has it gained widespread general acceptance?

Firefighters and incident commanders alike acknowledge that to enter a fully involved structure, the main body of fire must initially be knocked down, effectively slowing the overall momentum and intensity of the fire and thus minimizing any occurrence of flashover or other catastrophic event. Once the volume and intensity are reduced, advanced suppression evolutions can effectively continue. This method is generally reserved for specific applications where there is a large volume of fire emanating from a structure upon arrival, thus inhibiting any interior access until it is addressed. It is also reserved for those structures which are deemed unsafe to enter due to compromised structural integrity and a comprehensive risk versus reward analysis determining there is nothing to gain by placing personnel in harm's way. Any well-trained, disciplined, and experienced company or chief officer can make a sound decision based on the information at hand on whether to engage from the outside or to enter the structure for interior operations, including rescue. Removing any option of discretionary flexibility and reassessment of operations based solely on the no entry policy is a hindrance, not an advantage.

The most recent wildland fires in California were all conducted as exterior fire attacks, yet there still were fatalities and total devastation to the geographical area, all of which destroyed families and will cost billions to recover.[42] How is the exterior-only premise applied to these situations? Interestingly though, most fires are still fought with water, much like prehistoric days. As a seasoned company officer, I often directed my crews to poke an inspection hole into a warm wall to confirm the reading of the thermal imager, as well as obtain further information such as intensity, velocity, travel, and so on. If suppression crews are outside the building, none of this heads-up info can be obtained. Historically, this concept has been the most effective in formulating tactical suppression evolutions; no reinventing the wheel is needed. Although not ranked statistically in the top 10 of the Occupational Safety and Health Administration's most dangerous jobs in America,[43] unfortunately, the fire

---

42. "2022 Incident Archive," California Department of Forestry and Fire Protection, 2022, https://www.fire.ca.gov/incidents/2022.

43. Bureau of Labor Statistics, "National Census of Fatal Occupational Injuries in 2021," December 16, 2022, https://www.bls.gov/news.release/archives/cfoi_12162022.pdf.

service still has a high rate of injuries and fatalities because of the inherent dangers of the work, which is why we are constantly introducing improvements in tools, equipment, tactics, and gear. Firefighters have always maintained the ability to adapt and overcome, not simply throwing up our hands and deciding it's too dangerous for us. During the horrific February 2018 mass shooting at the Marjory Stoneman Douglas high school in Parkland, Florida, which left scores of young people dead and wounded, there was in fact an on-duty police officer at the school during the shooting. His subsequent ineffectiveness and unwillingness to engage based on his perceived information came with a high cost: lives lost![44] There are consequences for inaction or underaction in the public safety arena and limiting direct engagement in fire suppression evolutions because it's not safe is counterintuitive to our mission. The bottom line is simple—it doesn't take a Rhodes Scholar to comprehend that firefighting is inherently dangerous work. However, there are accepted risks fire departments recognize to fulfill our mission to save lives, stabilize incidents, and protect property. Just as soldiers fighting in a war, unfortunately, some get hurt or even die, which is why the services work so hard to minimize risk. The study of fire science has afforded our industry the chance to obtain the most accurate information. This research allows firefighters to formulate best practices and make sound decisions, acquire modern functional equipment, formulate comprehensive operating policies and standards, and achieve the best outcomes for our citizens.

Risk cannot be totally eliminated to save lives, stabilize incidents, and protect property; we strive to minimize inherent danger through comprehensive risk management programs, utilizing state-of-the-art tools and equipment, following best standards and practices, and incorporating consistent reevaluation for improvement. Career fire departments are sworn public employees charged with the duties to serve and protect, and are expected to deliver every time, not just when things are deemed safe! Unlike garbage pickup crews or other municipal employees who are not sworn to protect, firefighters and police take an oath upon being sworn in, as does the chief of department. A refusal to deliver on the promise to serve and protect is negligent, criminal, and simply goes against everything the fire service stands for. The chief of department, along with the governing state and local administrations, should be doing everything possible to minimize the risk while simultaneously affecting positive change to protect both firefighters and civilians, including best-practice policies. About 20 years ago, the national trend was to utilize plastic shopping bags at grocery

---

44. Audra Burch and Alan Blinder, "Parkland Officer Who Stayed Outside During Shooting Faces Criminal Charges," *New York Times*, June 4, 2019, https://www.nytimes.com/2019/06/04/us/parkland-scot-peterson.html.

stores to save the trees and forests from unnecessary deforestation. Today, the trend is back to paper bags because it turns out plastic bags are bad for the environment and wildlife. Here we are again, vacillating between trends because it seems like a good idea at the time. However, the feel-good trend backward to the good old days of past exterior-only fire attack makes for unsound practices based on a false narrative.[45]

Case Study

By now, I had been captain of Engine Company 10 for 5 years, but both my driver and senior member had been assigned there longer. On one (24-hour) tour my assigned driver was off, and it was my senior member's turn in rotation to be detailed as acting driver when around 4:00 a.m. we received a call first due for a building fire. He knew the district better than all of us because of the timespan assigned to the company, was an experienced and always competent firefighter and acting driver, and I had no issues with his performance to date. The address was several blocks to our north and was an easy route: turn right out of the firehouse and go three blocks and turn left; the building address was up one block on the next intersecting side street (the hydrant was right in front of the reported address). In Hartford, each apparatus adheres to specific routes when responding to facilitate the order of the assignment, the correlating hydrants, and the correct positioning approaching the fireground.[46] Deviation from your route is acceptable if your company is not in quarters and already on the road, which often happens; we adjust accordingly and communicate our approach to the incoming chief officer as well as all the other companies on the assignment. When the firehouse alert system announced this assignment, my (acting) driver quickly checked the map in the watchroom to confirm he was going to the right place and taking the closest hydrant, which is common practice especially for all drivers. He jumped into the driver's seat and careened out of the firehouse, first turning right, then inexplicably took the next left and pulled over in front of the first hydrant that appeared, parked the apparatus, and announced, "We're here!" as the parking brake made its characteristic clunking noise. Naturally, I turned to him in disbelief and *forcefully* questioned him (use your imagination) as to what he was doing. The bewildered stare he

---

45. Leigh H. Shapiro, "Interior Fire Attack: Obsolete or Indispensable?" *Firefighting* (blog), Fire Engineering, October 13, 2021, https://www.fireengineering.com/firefighting/interior-fire-attack-obsolete-or-indispensable/#gref.

46. Hartford Fire Department, *Administrative Manual—Department Directives*, 2015.

gave me looked as if there was something wrong with me! I quickly explained that this was not the address and we needed to hustle to catch up—after all, we were first due and there was no traffic on the roadway, so we should be arriving already. He quickly reengaged his driving duties as I guided him into the proper address; however, this time we arrived from the back way, meaning we came in from the wrong direction and parked in front of the address facing the wrong way. Once the incident was mitigated and while still on scene, I asked my driver to explain his actions as to what transpired. He said he believed he had done the right thing and only realized something was wrong when I initially questioned him. The chief officer in charge of the incident called me over to ask why my apparatus was facing the wrong way; he was aware we were not on the road at 4:00 a.m. returning from another call and there was no logical explanation for the first-due engine apparatus facing the wrong way in front of the building. I retorted jokingly, "Don't ask! We're here, that's all that matters right now!" He smiled, which indicated that after many years serving under him and building relationship capital, he trusted me to handle this situation and solve the problem. As soon as we returned to the firehouse (around 4:45 a.m.), I immediately pulled my acting driver into the office and questioned him: "What happened, what were you thinking, why did you do this? Is there something going on prior to this incident, either at work or at home, that would be a contributing factor impacting your behavior at such a critical time?" He really had no logical answer other than to say he had a mental malfunction, or a momentary break-down of his cognitive ability to process information and perform the required task. This didn't help me fix the issue because there was nothing concrete to work with, so I was compelled to figure this out on my own. He was under no more stress responding to this incident that any other incident because he was highly experienced in the acting driver capacity. I made suggestions like going to bed earlier to gain more rest, and possibly eating healthier. I gently offered both the Employee Assistance Program and the Chaplain Corps with the department receives services from, because often certain behaviors are a manifestation of a larger, looming problem. I also initiated a new guideline that when he drives, he will verbally acknowledge the address and repeat it back to me before we are en route, to ensure that he is responding to the proper address and to possibly clear up any cognitive speedbumps he may encounter. I told him I would monitor his future behavior and follow up during the next tour with some training regimens consisting of streets and routes. I then documented this episode fully and contacted the tour commander (who was also the IC on this incident) to follow up as to my corrective course of action. It was important to reach out and notify him that I in fact was doing something about what had transpired, even though it was a relatively innocuous occurrence.

I covered what may have had a contributing factor before this driving incident, what I did during the incident, and what I put forth after the incident to engage corrective actions and help prevent future occurrences. Now that I was fully aware of the possibility of this type of behavior occurring again, I instituted a plan of quickly swapping acting drivers (someone else in my crew would take over driving duties) if need be and alerting other companies via the radio that my company was experiencing a delay in response. I informed my other crew members that they may be called to drive in a pinch if such a scenario reoccurred which provided notice that they need to be better prepared. For this specific incident in the front seat of the apparatus responding to the building fire, things could have been significantly worse if not corrected immediately and effectively. Often, it's the human element that fails, not equipment or policy. As firefighters and officers, we need to be fully prepared to deal with the voids and gaps in our tactics and possess the willpower and fortitude to utilize our own initiative to complete the tasks at hand. My department's labor agreement states clearly in section 3.11 titled "Equitable Distribution of Work" that "the officer in charge shall apportion all . . . details among subordinates as equitably as practicable,"[47] so simply removing him from driving or denying him the opportunity would be a violation of our contract. Furthermore, as the officer in charge of both the company and its personnel, it's my duty and responsibility to utilize every available tool in my toolbox to correct any issues, provide guidance and training, and offer any available resources that may have a positive impact on performance. There's an old adage in the fire service that says, "It's not how you screw up, but how you recover that's most important."

Questions

1. As the officer, how would you proceed if your driver asserted that they did nothing wrong and was unwilling to take accountability for their actions?
2. What steps would you take if the IC demanded an explanation and insisted there be some type of correction, discipline, and training element?
3. How would you apply corrective action steps to other disciplinary issues you may encounter both in the firehouse and on the fireground?

---

47. "[Collective Bargaining] Agreement Between the City of Hartford and the Hartford Firefighters Association, July 1, 2009, through June 30, 2016," https://www.hartfordct.gov/files/assets/public/human-resources/hr-documents/hartford-fire-fighter-association-local-760-contract-7.1.2009-6.30.2016.pdf.

7

Analysis of an After-Action Report

As a deputy chief functioning as the incident commander (IC), I was often asked to complete a post-incident analysis in the format of an after-action report for significant incidents. At one particular incident, a working fire that occurred at 9 Carpenter Street, on August 23, 2015, a firefighter suffered minor injuries during suppression operations.[1] After conducting a thorough examination, it was determined to primarily be caused by the introduction of thermal hoods into the personal protective equipment (PPE) ensemble and the subsequent behaviors that followed their use. At 11:58 p.m., a box alarm assignment was dispatched at the aforementioned address for a reported structure fire; the assignment consisted of three engine companies, two ladder companies, a rescue company, and a chief officer.[2] Dispatch stated that fire was coming from an air conditioning unit on the second floor. On arrival, companies reported fire showing from the second-floor dormer on the Delta side of the structure and initiated an interior attack (fig. 7–1).

As the IC, I was not satisfied with the progress and transmitted a second alarm at 12:25 a.m., resulting in the dispatch of three additional engine companies, a ladder company, and an additional chief officer. I also revised my original incident action plan (IAP) and ordered the companies to transition to a defensive posture (fig. 7–2). The crews knocked back the main body of fire and immediately reentered the structure to complete extinguishment and overhaul. A comprehensive review of all aspects surrounding the incident revealed several key factors, categorized as strengths and weaknesses.

---

1. Hartford Fire Department, Post-Incident Analysis, 9 Carpenter St., 1-1-2 Second Alarm Working Fire, August 23, 2015.
2. Hartford Fire Department, Post-Incident Analysis.

FIG. 7–1. Crews were faced with heavy fire conditions on the second-floor rear while the first floor remained relatively clear. (Photo courtesy of Pat Dooley)

FIG. 7–2. Heavy fire conditions in conjunction with the failure of turnout gear ultimately routed interior crews and the order was given to go defensive. (Photo courtesy of Pat Dooley)

Strengths

- **Policies executed as designed:** The department's directives provided firm strategic and tactical guidance for structure fire operations and were adhered to by all personnel throughout the incident.
- **PPE ensemble performance:** One firefighter suffered only a minor burn to his finger from the initial (interior) operations. The firefighter's PPE ensemble sustained significant damage from the hostile conditions encountered: the turnout gear, fire and heat-resistant hood, helmet, and the self-contained breathing apparatus (SCBA) all received extensive thermal damage (figs. 7–3a, b, c, d, and e). If not for properly wearing the ensemble, the firefighter would have surely sustained significant injuries.
- **Command structure:** The department's command structure is designed to account for and protect all personnel operating at an incident. Another chief officer on scene and I ensured that accountability and oversight were in place throughout the duration of the incident, utilizing a risk analysis model to monitor progress and revise IAP to address rapidly changing fireground conditions. As warranted, the original IAP was modified, and companies were ordered to transition to a defensive posture. Crews methodically evacuated the structure and command conducted an immediate personnel accountability report (PAR) roll call.
- **Transitional strategy implementation:** Companies initially made an aggressive interior (offensive) attack to contain and control fire volume. The initial efforts, however, were ineffective in achieving knockdown. The revised plan to evacuate the structure and transition to a defensive operation facilitated the companies in quickly obtaining control of the fire. This was followed up with a subsequent secondary interior attack that proved adequate and successful.

Weaknesses

- **Personnel stressed from previous incident:** Most personnel at this incident operated at another working fire approximately 4 hours prior (10:28 p.m.). High ambient environmental temperatures and humid weather conditions impacted crew

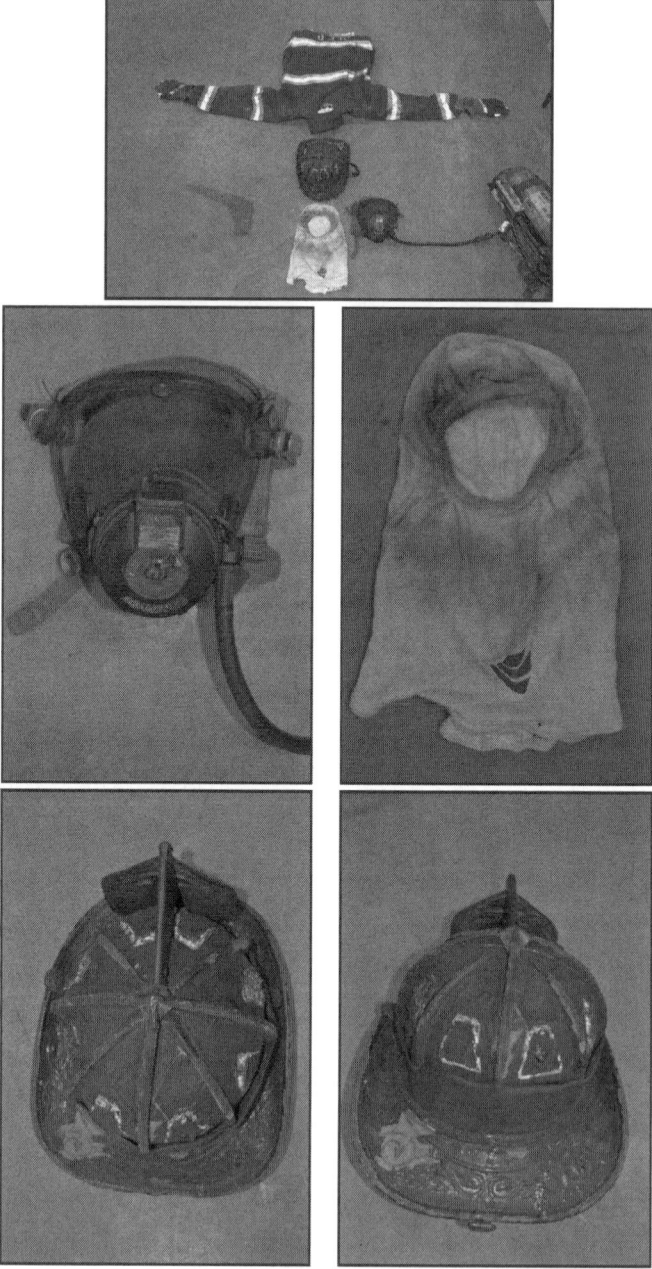

FIGS. 7–3a, b, c, d, and e. Part of the PPE ensemble worn by the injured firefighter. Note blistering of the SCBA regulator and facepiece and scorching of the fire and heat-resistant hood, as well as the thermal insult caused to the finish and the reflective material of the helmet. (Photos courtesy of the Hartford Fire Department [HFD])

endurance and, coupled with the previous strenuous activities, may have impacted crew effectiveness. Although the previous fire's magnitude was not significant enough to deploy the thermal hoods during suppression evolutions, the scope and duration of evolutions must be noted.

- **Water supply issues:** Two separate engine companies reported dead hydrants while initiating water supply operations, which forced companies to seek alternate hydrants and slowed suppression activities until a positive water supply was established. Additionally, the water mains in the area were all 8 inches or smaller in diameter, and the additional flow requirements needed for suppression operations proved to be inadequate.
- **Building construction features and configuration:** The structure was a legacy construction, wood-frame, one-and-a-half-story multiple dwelling with a gabled roof that had dimensional-lumber beams, rafters, and joists as structural elements. Based on city records, the building was constructed as a single-family dwelling in the 1920s and was converted into a multiple dwelling with units on the first and second floors sometime prior to 1999. Access to the second-floor unit was added by a narrow staircase within the enclosed front porch. The resulting staircase and second-floor landing travel path consisted of several acute pinch points. This access was used by the first-due companies and ultimately hindered the initial interior operations because of the location, position, and narrow dimensions of the stairwell causing a bottleneck for personnel and attack lines.
- **Attack line deployment:** The stairwell configuration and the resultant logjam condition that ensued played a direct role in the ineffectiveness of the initial attack line, yielding flow rate and advancement issues. The insufficient training, limited experience based on tenure, and situational awareness of these crews were also contributing factors to the situation.
- **Portable radio communications:** Hydrant personnel are often left at a hydrant some distance from the fireground and use portable radios to receive orders to open the hydrant. Similarly, the pump operators are instructed to charge attack lines via the company officer's portable radio. Often, personnel can become anxious and use the portable radio to ask if the pump operator or company officer would like the supply or attack line charged. Additionally, transmissions from second-alarm companies setting

up exterior operations serve only to compound the situation. The unnecessary radio traffic that was present at this incident tied up the incident channel, preventing critical messages from being transmitted. Although reliable communication is a key to effectively and safely managing emergency incidents, ill-timed or unnecessary transmissions impede ICs from receiving and acknowledging Maydays, urgent transmissions, updates from interior crews, and other critical messages regarding the changing conditions.

- **Recent addition of hoods to PPE ensemble:** Although the proper wearing of the complete PPE ensemble at this incident prevented at least one firefighter from receiving serious thermal injuries, further analysis and personal testimony of the crew involved, however, indicated the addition of the fire and heat-resistant hood led to the actions leading up to the event and caused the injury. The department added these fire- and heat-resistant hoods to the required PPE ensemble on July 27, 2015, just one month prior to the incident.[3] Personnel received classroom instruction and live-fire training in a controlled environment prior to implementation of the mandate. The Carpenter Street incident was the first working fire in which the hoods were utilized on a department-wide basis. Prior to this fire, personnel relied heavily on their exposed ears and neck to determine if conditions warranted evacuation or whether to push farther into a structure during interior operations. Eliminating this sensory indicator requires personnel to utilize other sensory input and indicators for evaluating interior conditions and maintaining situational awareness. At this incident, fire conditions had begun to thermally degrade the PPE ensemble before the (injured) firefighter was able to recognize and react to the rapidly deteriorating conditions (fig. 7–4). Crews must use thermal imaging cameras and read the smoke conditions to monitor the environment and the fire conditions and maintain an appropriate level of situational awareness. Consciously utilizing this process, however, takes extensive training and discipline, especially for less-experienced personnel.

---

3. Hartford Fire Department, *Administrative Manual—Department Directives,* 2015.

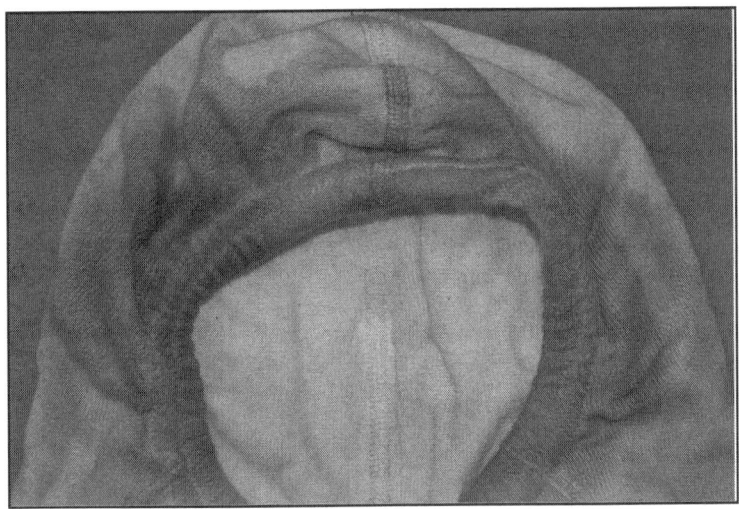

FIG. 7–4. Close up of the scorching at the right temple area of the fire and heat-resistant hood (Photo courtesy of the HFD)

As the IC, I ordered all companies to evacuate the building to initiate a (temporary) defensive attack to knock back the main body of fire from the exterior. As the PAR was concluding, the injured firefighter and his crew came up to me at the command board to verbally report their safe evacuation from the building. I looked over at the injured firefighter, who looked completely disheveled: His turnout coat was smoking, and he was patting himself with his gloved hand as if to snuff out smoldering pieces of his turnout coat, creating puffs of wispy smoke (fig. 7–5). I asked with bewilderment, "What the hell happened to you?" He responded with a description of the events that had just unfolded. Once reaching the top of the narrow stairwell that was used for the initial attack, the crew encountered moderate smoke and a glow of fire at the second-floor landing. Because the crew could not accurately read the thermal conditions within the environment, they pushed forward toward the glow because they were searching for the seat of the fire and not merely attempting to flow water on a visual illumination. On doing so, they immediately encountered an obstacle, a pool table, at the top of the stairs in the front room. Seeing no other way to enter this dwelling and get a handle on the fire, the injured firefighter decided to crawl across the top of the pool table to advance the line, placing himself even higher up in the superheated environment. Once they were beyond the obstacle, and with the fire free-burning, operating temperatures quickly rose beyond the limits of their turnout gear and the single 1¾" handline. Within seconds, the firefighter realized his gear was rapidly failing and an emergency evacuation was required. The extremely narrow staircase

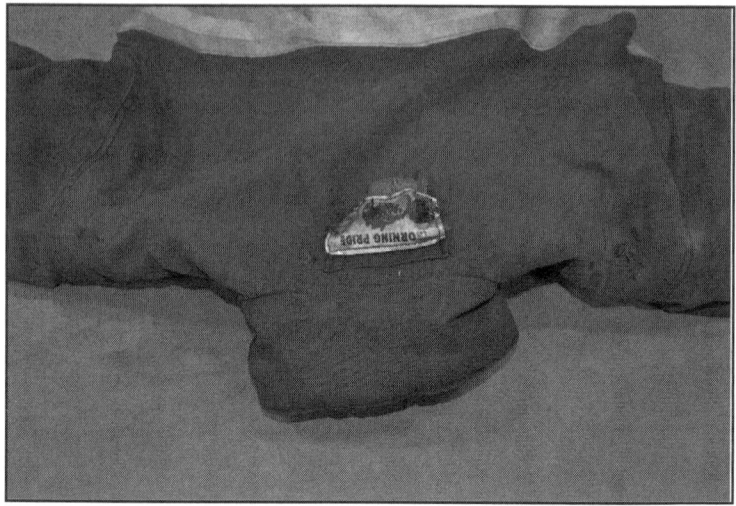

FIG. 7–5. Note the thermal degradation present on the shoulders of the turnout coat. (Photo courtesy of the HFD)

hindered the quick deployment of a second backup line, which further contributed to this situation. Although the crew safely evacuated without delay, the injured firefighter still suffered a minor burn to one finger where his glove had failed.

Fire departments must continue to improve their delivery of services and ability to maintain safety through consistent, realistic, aggressive training evolutions and education programs. As a less-experienced workforce becomes the trend resulting from a reduction in fire duty and attrition, this becomes even more critical. Coupled with a stringent accountability system, these actions can improve the quality of service provided and reaffirm a commitment to public and firefighter safety. A cultural paradigm shift centered on high performance is the fundamental key to individual and organizational improvement. Although technological advances and the addition of new equipment can be a tremendous benefit to safety and performance, their limitations and any necessary precautions must be fully identified and understood. Fireground failures and operational inadequacies can be addressed through such proactive measures, but there is no substitute for experience. Organizations must place value on fireground experience and ensure that it is shared and preserved for generations to come.[4] The point here is not to villainize hoods, but to illuminate the

---

4. Leigh H. Shapiro, "Encapsulation of Firefighter Illustrates Need for Critical Thinking Skills," *Leadership* (blog), Fire Engineering, March 1, 2020, https://www.fireengineering.com/leadership/encapsulation-of-firefighter-illustrates-need-for-critical-thinking-skills/#gref.

deficiencies that may arise when training and experience do not match with the introduction of new, critical equipment, and how fireground failures, although in this case not fatal, can be caused by these elements. It's not so much a matter of what happened, but more importantly why it happened, and it helps us to understand how to shape future policies and procedures.

Case Study

Every department strives for continuous and timely development and adoption of improved services and equipment while maintaining safety. Unfortunately, departments are often consumed with sustaining their mandated training benchmarks, leaving some skillsets by the wayside. The understanding is clear: modern fire departments must deliver more while having less. There is a positive trend in the reduction of fires and fire duty mostly attributed to such factors as the availability of cellphones to rapidly report a fire, thus minimizing an incident, and more stringent fire safety codes. However, less fire duty directly correlates to a less-experienced suppression workforce, thus compelling departments to find new ways to maintain and advance skillsets, accountability, and ultimately a culture of high performance. The rapid advancement and adoption of new technology and equipment helps drive performance, but with all new and improved resources, the fire service still must maintain their old school ways of getting the job done. The reality is that we are compelled to understand and appreciate the limitations and boundaries of progressive modernization. That is why embracing experience from wherever you may find it is so relevant and sought out today. To assist in creating a well-rounded, prepared, and ready-for-work firefighter, training above and beyond department mandates has become popular. However, with limited budgets and staffing, such experiential training often must be garnered from outside the confines of the department.

Questions

1. How do you engage the newer firefighters to learn from sources other than the department's training division?
2. How would you integrate the senior personnel with the younger, less-experienced personnel to establish a solid foundation and environment of constant and consistent learning and mentoring?
3. How would gauge the progress and degree of absorbed learning those mentors provide to your less-experienced personnel?

8

The Incident: The Impact of a Fire Service Line-of-Duty Death

It is my aspiration to never have this type of incident occur ever again anywhere within the U.S. Fire Service (fig. 8–1). But unfortunately, it still does happen with unwanted frequency, so we must learn from it and pass along that knowledge to others. In the 1974 blockbuster movie *The Towering Inferno*, Battalion Chief (BC) Mike O'Hallorhan, aptly played by movie star Steve McQueen, confronts the owner and operator of the high-rise building played by actor William Holden, who in that scene is hosting a gala in the grand promenade ballroom on the top floor of this megastructure.[1] When Chief O'Hallorhan informs Holden that he should announce that there is a

FIG. 8–1. In remembrance of those who made the ultimate sacrifice (Photo by Pat Dooley)

---

1. *The Towering Inferno*, directed by John Guillermin (20th Century Fox, 1974).

fire in the building and the party should be moved below the fire floors to an area not affected by the fire, Holden asks, "Just how bad is it?!" to which an unphased, seasoned BC unemotionally quips, "It's a fire mister, and all fires are bad!" (fig. 8–2).

FIG. 8–2. Chief Mike O'Hallorhan (Screenshot of video clip courtesy of 20th Century Fox)

This deadpan delivery of a rebuke to the innocuous question of "How bad is it?" smacks with years of experience one would expect from a senior chief officer. All fires are bad. Some are worse than others; however, fires that result in the loss of life, especially firefighter loss of life, are no doubt as bad as it can get.

The Ultimate Sacrifice

Unfortunately for me, I was the assigned safety officer at a house fire on October 7, 2014, which resulted in the loss of one of our firefighters, Kevin L. Bell, in Hartford,[2] which was referenced in the beginning of this book. Up until then, my department and city had not experienced a line-of-duty death (LODD) since 1974 when a collapsing brick garage building claimed the life of Firefighter (FF) Thomas A. Fischer.[3] Nobody expected to be a part of a LODD incident, certainly not me. My department, like most fire departments, was not prepared for this event. Unfortunately, bad things happen, and in our line of work, there is only so much preparation that can be done to manage this type of event. All

---

2. Hartford Fire Department Firefighter Kevin Bell line-of-duty death while operating at 598 Blue Hills Ave., October 7, 2014.
3. Hartford Fire Department, Firefighter Thomas A. Fischer line-of-duty death while operating at Box 425, September 15, 1974.

involved learned some hard lessons that night, and there is inherent value in sharing experiences from that fateful event and beyond. This review is in no way meant to disparage, blame, or make light of what transpired; it simply aims to illuminate how events transpired and only speaking directly of certain elements of the incident and what happened with whom. For further details, you will have to read all official reports that were generated on this incident. Just because we do everything by the book and have strong policies and the latest equipment doesn't mean we can completely prevent something bad from happening. As author Fran Liebowitz aptly put it, "Your bad habits can kill you, but your good habits won't save you!"[4]

The Aftermath

Dealing with the trauma of the incident was surreal for all involved. The morning following the fire, our local newspaper the Hartford Courant had a small blip of a story on the bottom of the front-page fold. The next day it was totally front-page news, as depicted in the photograph seen in figure 8–3.[5]

FIG. 8–3. These editions of the Hartford Courant are from October 8th, the day after the fire, and October 9th, two days after the fire.

---

4. Fran Liebowitz, in *Pretend It's a City* (Netflix, 2021).
5. *Hartford Courant*, October 8 and 9, 2014.

Firefighters are generally used to seeing their handiwork on the front page of the local newspaper, if not somewhere within. This event was very different and shed a cold light on the tragedy that can occur within the fire service and to our chosen family members. The entire department was impacted, not just those brave firefighters who were operating at the scene that evening. Imagine being off duty and assigned to a different end of the city and walking into your firehouse the next morning to report for duty. The air is thick with emotion, tension, and bewilderment. The phones are curiously chiming among all the firehouses throughout the city to get a better perspective on the facts of who was involved, what happened, and why. Rumors are flying, the news is abuzz with information (some factual and some conjecture and inuendo), and the overall mood within the department is stunned and subdued. Not our proudest moment; no victory laps or accolades bestowed on the city's bravest. Just empathy, sympathy, and the simmering yet growing rumble for the demand for answers and accountability! As stated in the beginning of this book, FF Kevin Bell was subsequently located within the fire building and brought outside to an awaiting ambulance crew. I draw on the photograph taken of the Boston Firefighters whisking away one of their injured brothers at the March 26, 2014, Beacon Street fire in Boston's Back Bay, which ultimately claimed the lives of two of their bravest (fig. 8–4).[6] Take a good look at the faces of those firefighters desperately working to save their brother's life. With this fire, as with my LODD incident, whether you directly participated in the rescue and removal of an injured fellow firefighter, or you witnessed it while working on the fireground, you most definitely are in a state of shock.

The trauma of the visual grasps you in a profound way, much differently than if you were involved in a fatal civilian incident during the normal course of your duties. Seeing a civilian victim in cardiac arrest or severely injured

FIG. 8–4. The expressions of these firefighters cannot be expressed through words. (Photo courtesy of the Boston Globe)

---

6. "Back Bay Fire," *Boston Globe*, 2014, https://www.bostonglobe.com/2014/03/26/photos-back-bay-fire/HR8Mz7ZOk9XGVMf1Mv1HEO/story.html.

is significantly different than the visual of a firefighter in a similar physical state. We serve the public and are sworn to protect them, no matter what the incident circumstances are, but our fellow firefighters are not only considered peers and coworkers, but to many, they are members of our family as well. To see our people in that state of physical emergency is a total shock to the human nervous system, much like if a member of our own family was in that unfortunate perilous state. As for the incident commander (IC) of an incident where one or more firefighters are severely injured or worse, there most likely will be a temporary breakdown of operations because although we plan and train for such events, it is common for us to be caught off guard and not prepared to actually engage in this situation. This is primarily because the human reaction is first that of disbelief and disconnect in fully comprehending what is actually transpiring, therefore thrusting the IC into a situation where they have lost their initiative and now must directly deal with the rescue and removal of a firefighter. Although we train for a Mayday event, the visual of a downed firefighter can overwhelm your emotions and create chaos. Larger departments that have had some experience with this are better prepared for the event; however, smaller departments with little or no experience can be expectedly jolted and stunned when something like this occurs. Yes, we are prepared and ready, but when it occurs for us, there is always an inevitable lag in cognition.

For my LODD incident, the question that hovered over me for a long time was why: why did this happen, and selfishly, why to me? I always believed I operated in the safest manner possible, yet nobody expects these bad things to happen. The fire service is prepared with safety equipment, policies, and technology. We are also ready, with training and practice for most situations that keeps us sharp. Yet, ultimately, it may in fact come down to a matter of control: although we minimize the risks as much as possible, some things simply are beyond our control. I cannot control with absolute certainty what happens inside the fire building. I can only trust the officers and firefighters to perform to the standard of training and equipment they bring inside with them (figs. 8–5a and b).

Evidence Collection

One of the details of this tragic event that we clearly were not expecting or prepared for was the evidence collection. The format in which the investigation was conducted was two-pronged: our fire marshal's office would investigate the origin and cause of the fire and issue their report, and the state police would

FIGS. 8–5a and b. An interior view of the fire room where the LODD occurred and the wall where the second injured firefighter was using his ungloved hand to feel for his way out (Photo courtesy of the National Institute for Occupational Safety and Health [NIOSH])

investigate the LODD as an untimely death, and then issue their report as well (figs. 8–6a, b, and c).[7]

Everything the state police could get their hands on was seized as evidence to assist in determining any mitigating factors that would cause this LODD, beyond the obvious. Local TV news channel WTNH News 8 ran a clip of a Hartford police officer seizing the turnout gear and self-contained breathing apparatus (SCBA) of FF Bell and placing it in his trunk to be brought to the

---

7. Hartford Fire Department, *Administrative Manual—Department Directives*, 2015.

8 • The Incident: The Impact of a Fire Service Line-of-Duty Death   183

FIGS. 8–6a, b, and c. The symbols in patch form on the uniforms of the fire investigators

Connecticut Forensic Science Laboratory for inspection and testing (figs. 8–7a and b).[8] The narrative of the accompanying video is quoted here:

> Tonight, the Hartford Fire Department and the state police fire marshal's office are not officially commenting on the investigation into the death of Firefighter Bell. On the night of the fire, Hartford Police did take Bell's gear into custody and examine it to make sure it was working properly. News 8 has learned that the firefighter was low on air while fighting the fire.

---

8. "Questions over Deadly Fire," WTNH News 8, October 2014, https://www.youtube.com/watch?v=ROemjIm0Heo.

This was another gut punch for us when we witnessed it on scene. We all didn't go to work that day and expect to have our scene, equipment, and records seized as evidence for any reason. It placed a pall over all the firefighters involved because we were now subjects of an official police investigation. Firefighters are not used to being on this side of a situation. We are the ones responding and helping, not the ones being investigated by the police.

The evidence that was eventually collected by the police was voluminous and comprehensive (fig. 8–8). The first-due apparatus engine to which FF Bell was assigned and ultimately stretched an attack line off was seized for mechanical inspection to investigate if the apparatus malfunctioned in any way which may have had a factor in the incident, as well as all related maintenance records.[9] As stated earlier, both FF Bell's turnout gear and SCBA were seized for inspecting and testing to determine if the gear failed or the SCBA malfunctioned and contributed to his death.[10] His portable radio that was issued to him during his shift for his apparatus riding position was seized and inspected to investigate

FIGS. 8–7a and b. A Hartford police officer seizes as evidence the turnout gear and SCBA worn by FF Kevin Bell. The turnout gear and SCBA were sent to the Connecticut Forensic Science Laboratory to be inspected and tested. (Screenshots of video clip courtesy of WTNH News 8)

---

9. Connecticut State Police, Case Number CFS1400630622, October 9, 2014.
10. Connecticut State Police, CFS1400630622.

FIG. 8–8. Connecticut State Police Fire and Explosion Investigation Unit (CSP FEIU) utility truck parked outside the fire building serving as an on-scene mobile investigation command center. The fire building is on the left, secured by Hartford Police. (Photo courtesy of NIOSH)

for proper operation.[11] All audio recordings from the dispatch center were collected and carefully analyzed.[12] All the training records for each of the personnel on the first-alarm assignment were ultimately seized at our training division.[13] Those records would be thoroughly scrutinized to find any discrepancy or gaps in mandatory training that may have had a contributing factor.

All the year-to-date in-house interdepartmental correspondence at the quarters of Engine 16 where FF Bell was assigned was seized to determine if there were any ongoing problems or significant communications with equipment, personnel, and policy which may had a contribution to the incident.[14] All records in the firehouse and at the fire marshal's office regarding building inspections, area survey forms, and fire prevention inspections for the address of the fire were carefully reviewed for any pertinent facts.[15] Each one of the first-alarm on-scene firefighters, myself included, were interviewed in the days following the incident by the state police and compelled to give a sworn statement as to what we did and what we saw that night at the incident (figs. 8–9a, b, and c).[16]

---

11. Connecticut State Police, CFS1400630622.
12. Connecticut State Police, CFS1400630622.
13. Connecticut State Police, CFS1400630622.
14. Connecticut State Police, CFS1400630622.
15. Connecticut State Police, CFS1400630622.
16. Connecticut State Police, CFS1400630622.

CELL: (860) 523-5077

STATE OF CONNECTICUT
DEPARTMENT OF PUBLIC SAFETY
DIVISION OF STATE POLICE

Case Number: CFS1400630622
Date: 10-08-14
Time Started: 2200
Time Ended: 0030

WITNESS STATEMENT OF SHAPIRO, LEIGH H.

I, LEIGH H. SHAPIRO date of birth. 08-15-66
of 18 Nursery Drive, town/city of West Hartford, CT 06117

make the following statement without fear, threat, or promise. I have been advised that any statement(s) made herein which I do not believe to be true, and which statement is intended to mislead a public servant in the performance of his/her official function, is a crime under C.G.S. section 53a-157.

My name is Leigh H. Shapiro. I have been a member of the Hartford Fire Department and have been for 26 years. I am currently assigned as District Chief (DC 1) as the senior tour commander assigned to Tour A. I was working on 10-07-14. I am stationed at the Pearl Street station and have administrative responsibility for running the fire operations of the city but I also respond to calls. I cover the south end of the city. I came on duty at 0700 hours. It was a very busy day with a variety of serious calls, which is unusual. There was a suicide, a drowning, and a small fire on Maple Street. There were many other calls, too. That evening we had dinner and had just finished when the bell hit for the call on Blue Hills Avenue; I think it came in as smoke in the building. The guys on the Tac got up and left and I went to the office and listened to the radio. I could hear the guys talking on the radio about smoke and I put the radio back down on the desk because it was a desk radio and ran downstairs. My driver, a detailed man from Ladder 6, Francisco Lopez, was already in the car and I asked him if they called it yet. He told me that they had just called a 10-3 (working fire) and I said "ok, let's go." I could hear them talking on the radio about a working fire on the Dispatch Channel. I turned my portable onto the Incident channel and I could hear chatter. I remember that when I was northbound on Blue Hills Avenue going past Westborn Parkway I could hear JR (from Engine 16) calling his boss on the radio asking him if he wanted water. He must have asked like 4 times. This is not unusual for him to say, but all I could hear in response was a crackle on the radio and a very faint voice. It was virtually inaudible; it could have been anyone keying the mic, though, but I never heard a clear response. We have had issues with radio communications in this area before, so I wasn't sure if that was why the responses were inaudible. I rolled in right behind a single engine company. I was just a little ways behind them. When I arrived we parked directly behind district 2 and all the other companies were already there, except Engine 5. I got out of the car and I put my gear on, along with my driver. I grabbed my flashlight and walked over to DC 2 (Jim McLoughlin) who was standing in the sidewalk in front of the building and told him I was here. I was acting as the Safety Officer at this call and McLoughlin was the Incident Commander. He acknowledged me and I looked down the driveway on the right side and saw that there was no fire. I walked along the sidewalk in the front of the building and looked to see what was going on. I saw several hose lines going into the building through the front door. I walked down the left side of the building (B Bravo side) to the rear and looked up and could see the fire was venting out of the B/C corner windows and the attic. It was

By affixing my signature to this statement, I acknowledge that I have read it or have had it read to me and it is true to the best of my knowledge and belief.

Signature: X

CONNECTICUT STATE POLICE
Fire & Explosion Investigation Unit

Trooper Paul G. Makuc #885
Accelerant Detection K-9 Baxter #9047

269 Maxim Road
Hartford, CT 06114
paul.makuc@ct.gov

Cell: (860) 250-7672
Office: (860) 706-5600
Fax: (860) 706-5619

tatement and made oath before me to the truth of the matters contained

kuc #885 CSP-FEIU

Page 1 of

FIG. 8–9a. This is the actual statement I gave to the Connecticut State Police the next evening after the incident. It took 3 hours to compile, and the interview was conducted in the front seat of a CSP FEIU pickup truck. This speaks to the nature and immediacy of gathering relevant information. Page 1 of 3.

8 • The Incident: The Impact of a Fire Service Line-of-Duty Death

STATE OF CONNECTICUT
DEPARTMENT OF PUBLIC SAFETY
DIVISION OF STATE POLICE

Case Number: CFS1400630622

WITNESS STATEMENT OF: SHAPIRO, LEIGH H.

active fire in a free burn stage with not a lot of pressure pushing it out. I looked at the rear of the building and which completed my 360 observations. I determined that the fire was in the second floor corner room. I looked over at the exposure building adjacent to the fire building and after careful examination determined it was not involved. While I was doing my assessment observing the fire, I could see DC 2 standing at the A/B corner by the fence post. I yelled up to him "Where the fuck is 16's, why aren't they making the push?" He said something back but I don't remember what he said but the answer wasn't to my satisfaction. I walked up to where he was standing in the front of the building and I looked at the A side, top floor right side window and I saw "black fire," thick heavy smoke starting to billow out, which is indicative of a flashover. I didn't like the looks of that at all and it triggered my decision and I told him, "Get them the fuck out of there. Everybody out. Dump the building." I said this because the fire was still in free burn and it didn't appear that our operations were having any effect on the fire. I didn't like what I saw and decided it was time to reassess and change the plan. This was the safest thing to do. DC 2 got on the radio and ordered all companies out of the building and began doing a PAR. Simultaneously I saw a firefighter appear in the second floor window, the left hand window of the two on the right side. The window was partially open and I could see a firefighter's hand coming out the bottom open portion of the window and I could see the window starting to break from the inside by the firefighter trying to break it. Everyone started screaming to get a ladder and just as a ladder was being thrown up against the window the firefighter ejected himself from the leftmost window of the two and he fell to the ground below. I saw him come out of the window but I did not see him land on the ground. I noticed that when he came out the window his head was bare; he had no mask or helmet on, or so it appeared to me. The top of his head was all pink and sooted up. Several firefighters went to tend to him and other guys were coming out the front door from the building. I was focusing on everyone else getting out of the building. At that time I turned to the two aides and ordered a second alarm which was done immediately. I was standing with DC 2 in the front of the building when Lt. Moree came up to him and said, "I can't find Kevin Bell." The PAR was still going on. I said to Lt. Moree, "Where the fuck is he? What do you mean you can't find him? Where did you leave him?" Lt. Moree said he didn't know where he was at this time, but the last time he saw him was at the top of the stairs. I said "Are you sure? Are you sure he's not out here somewhere" and he said, "I don't know." DC 2 immediately turned to a crew on his right side and told them to take a line and get up there and find him. They did. They went through the front door and I followed them to the front door opening and looked up the stairs. They went up the stairs and they were up there for about 30 seconds and said, "We got him, we got him." It looked like they were struggling to get him out. Lt. Turner, the Lieutenant on Ladder 4 was up there screaming at them to get him out. They eventually got him free from whatever he was hung up on and they pulled him out down the stairs. I backed out onto the front walk and they laid

By affixing my signature to this statement, I acknowledge that I have read it or have had it read to me and it is true to the best of my knowledge and belief.

Witness: _____ Signature: X _____

Witness: _____

Personally appeared the signer of the foregoing statement and made oath before me to the truth of the matters contained therein.

I notarized, endorse here: _____

Detective Paul G. Makuc #885 CSP-FEIU

Page 2 of

FIG. 8–9b. Page 2 of 3.

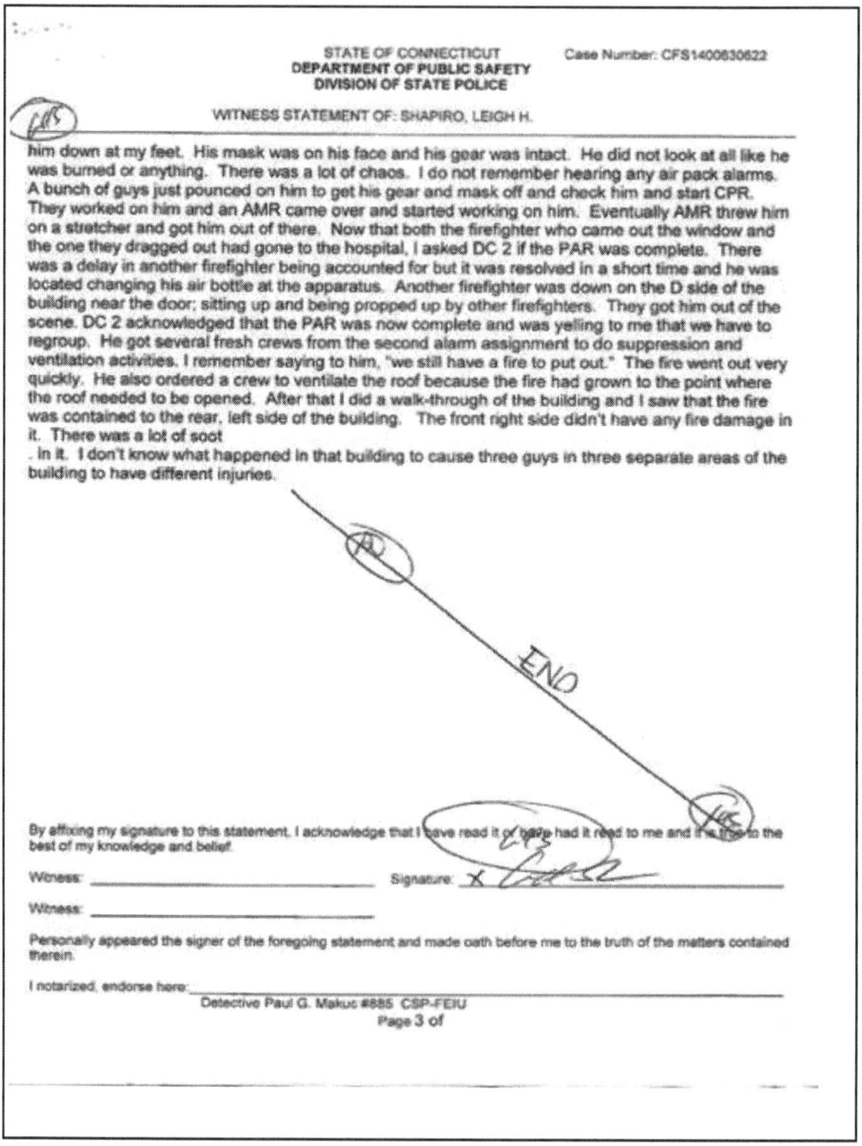

> STATE OF CONNECTICUT
> DEPARTMENT OF PUBLIC SAFETY
> DIVISION OF STATE POLICE
>
> Case Number: CFS1400630622
>
> WITNESS STATEMENT OF: SHAPIRO, LEIGH H.
>
> him down at my feet. His mask was on his face and his gear was intact. He did not look at all like he was burned or anything. There was a lot of chaos. I do not remember hearing any air pack alarms. A bunch of guys just pounced on him to get his gear and mask off and check him and start CPR. They worked on him and an AMR came over and started working on him. Eventually AMR threw him on a stretcher and got him out of there. Now that both the firefighter who came out the window and the one they dragged out had gone to the hospital, I asked DC 2 if the PAR was complete. There was a delay in another firefighter being accounted for but it was resolved in a short time and he was located changing his air bottle at the apparatus. Another firefighter was down on the D side of the building near the door; sitting up and being propped up by other firefighters. They got him out of the scene. DC 2 acknowledged that the PAR was now complete and was yelling to me that we have to regroup. He got several fresh crews from the second alarm assignment to do suppression and ventilation activities. I remember saying to him, "we still have a fire to put out." The fire went out very quickly. He also ordered a crew to ventilate the roof because the fire had grown to the point where the roof needed to be opened. After that I did a walk-through of the building and I saw that the fire was contained to the rear, left side of the building. The front right side didn't have any fire damage in it. There was a lot of soot in it. I don't know what happened in that building to cause three guys in three separate areas of the building to have different injuries.

FIG. 8–9c. Page 3 of 3.

Chapter 25 of the *National Fire Protection Association (NFPA) 921: Guide for Fire and Explosion Investigations* spells out the specific protocols for investigating fire and explosion deaths and injuries, and this information was adhered to when the investigations were conducted.[17] The Hartford Fire Department formed a board of inquiry, and after extensive interviews with personnel from the entire first-alarm assignment, they issued their report (fig. 8–10).[18] All those personnel were subsequently interviewed by a NIOSH team of investigators, and they issued the official NIOSH report some months later (fig. 8–11).[19] The CSP FEIU issued their final report revealing their findings.[20] The insurance carrier for the property owner also issued a report as to their findings, and the Occupational Safety and Health Administration subjected the fire department and the city to numerous fines based on several critical policy and procedural gaps and failures which resulted in the injuries of those operating on scene, as

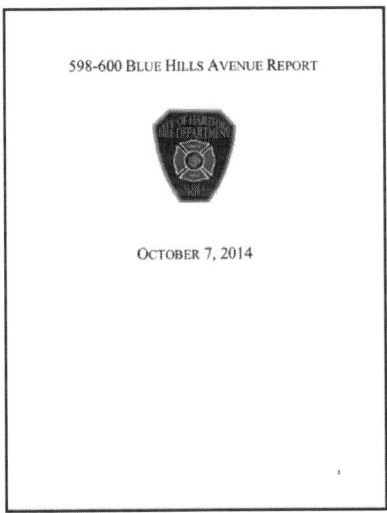

FIG. 8-10. HFD Board of Inquiry report

---

17. *NFPA 921: Guide for Fire and Explosion Investigations* (Quincy, MA: NFPA, 2021).
18. Hartford Fire Department, Board of Inquiry Report on the Fatal Fire of October 7, 2014, August 7, 2015.
19. NIOSH, "Career Firefighter Dies from an Out of Air Emergency in an Apartment Building Fire-Connecticut," Fire Fighter Investigation and Prevention Program F2014-19, January 24, 2017, https://www.cdc.gov/niosh/fire/pdfs/face201419.pdf.
20. Connecticut State Police, CFS1400630622.

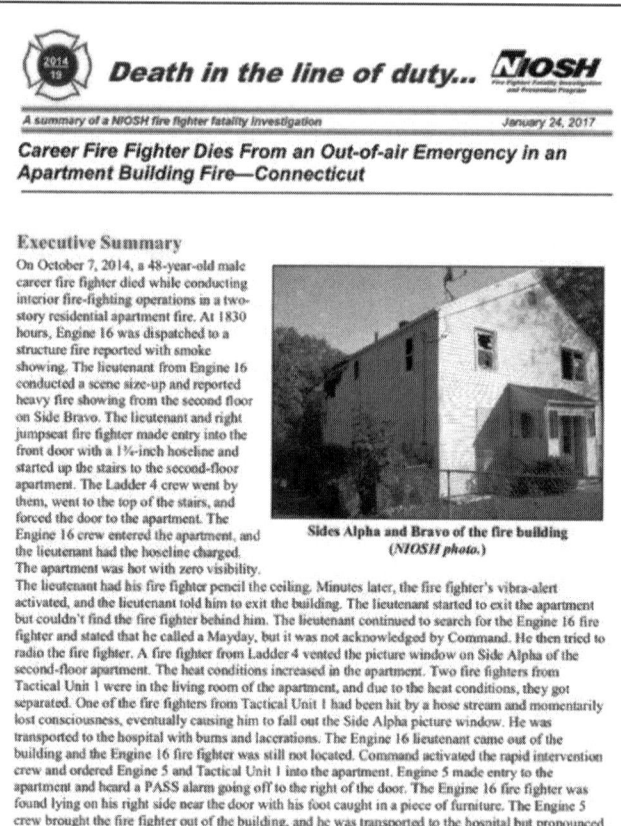

FIG. 8–11. NIOSH LODD report (Photo courtesy of NIOSH)

well as the death of FF Bell.[21] Specific protocols were followed for the process of investigating the untimely death of a firefighter. The National Fallen Firefighters Foundation (NFFF) has developed a comprehensive resource guide for the handling of firefighter LODDs (fig. 8–12).[22] In it are topics such as pre-incident planning for assisting families in their time of need, proper

---

21. Occupational Safety and Health Administration, Inspection 999065.015, October 7, 2014.

22. NFFF, "A Resource Guide for Handling Firefighter Line-of-Duty Deaths," National Interagency Fire Center, https://www.nifc.gov/sites/default/files/pio/documents/ResourceGuideHandlingFF_LoDD.pdf.

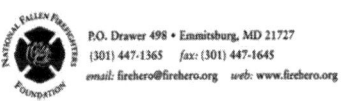

FIG. 8–12. This is the guide which was utilized to assist the department in handling the aftermath of this incident. (Photo courtesy of the NFFF)

methods of notification and support of family members, and overall department support. This guide is an invaluable resource for any department in their time of need, and after receiving numerous offers for assistance by other fire departments who had experience in these matters, my department followed these protocols closely.

The U.S. Fire Administration has created a Firefighter Autopsy Protocol that was strictly followed by the State of Connecticut Office of the Medical Examiner immediately following the incident (fig. 8–13).[23] In it, there are specific protocols regarding manner, cause, and mechanism of death; evidence collection and documentation; and an exhaustive list of occupational aspects of firefighting of specific concern to autopsy. Included is a thorough examination of all the personal protective equipment the decedent was wearing during the event.

---

23. Jeffrey O. Stull and the U.S. Fire Administration, "Firefighter Autopsy Protocol," International Personnel Protection Inc., March 2008, https://www.usfa.fema.gov/downloads/pdf/publications/firefighter_autopsy_protocol.pdf.

FIG. 8–13. Up until this incident, I was unaware that this protocol existed. Specific elements of evidence are gathered to create the relevant report indicating exactly what the cause of death was. (Photo courtesy of the U.S. Fire Administration)

The International Association of Fire Chiefs (IAFC) has developed a comprehensive protocol concerning the notification of surviving family in the event of a LODD of a firefighter (fig. 8–14).[24] This was also followed for our event. The Connecticut Statewide Honor Guard has developed their own incident action plan in the event of a LODD incident and has engaged this plan several times since this incident.[25] Incorporated in it are all the specific protocols which a department can draw from as a resource and guide.

The Unexpected

Several aftereffects immediately transpired from this deadly fire incident in which we lost one of our brothers (fig. 8–15).

---

24. "ICHIEFS LODD Response Plan: Five Principles of Notification," International Association of Fire Chiefs, 2018, https://www.firehero.org/resources/department-resources/sops/.
25. Connecticut Statewide Honor Guard, "Line-of-Duty-Death Incident Action Plan," March 2011, https://www.firehero.org/wp-content/uploads/2014/05/ConnecticutStatewideLODDIAP.pdf.

ICHIEFS

LODD Response Plan

Five Principles of Notification

In Person

- Always do the notification in person, never by phone!
- For family members living out of the local area, arrange for authorities in that area to make the notification in person.
- Immediately find the firefighter's emergency contact information to know who needs to be notified in person. Usually, the spouse (or unmarried partner) and parents of the firefighter should be the first priority.

In Time and with Certainty

- Before making notification, have positive identification of the deceased firefighter. Obviously, errors in identity can cause extreme trauma.
- Never discuss a fatality over the radio. This may result in a family member receiving the news before you can notify them in person.
- Quickly gather as much information about the incident as possible before making the notification. Survivors will likely have questions.
- Get to the survivors quickly. Don't let the media notify them first.

In Pairs

- Have two people present to make the notification. Survivors may experience severe emotional or physical reactions when they learn of the death.
- Use the employee's emergency contact information to identify a uniformed member of the fire service to accompany the department's representative. It is helpful to have the department chaplain or a friend of the family, too.
- Take two vehicles, if possible. This will allow one of you to take a survivor to the hospital, if necessary, while the second person stays with other survivors.
- Before you arrive, decide who will speak and what that person will say.
- Rehearse what you are going to say. Consider writing it down.

FIG. 8–14. Even the way in which notifications are conducted is spelled out in a protocol. Not anything like you see in movies—this is serious business and must be handled with dignity, respect, sensitivity, and honor. (Photo courtesy of the IAFC)

FIG. 8-15. This plaque and flag hang within the quarters of Engine Company 16 where FF Bell was assigned and ultimately lost his life. (Photo courtesy of Pat Dooley)

Several details came to light after this fire incident occurred. First, in the ensuing chaos of removing FF Bell to an awaiting ambulance to be rushed to the emergency room, one of the civilian photographers in our official photo unit jumped into the back of the ambulance and began administering cardiopulmonary resuscitation compressions on FF Bell. Within our commercial ambulance service contracted with the city, each unit is manned by one paramedic and an emergency medical technician (EMT). The medic was intensely working on FF Bell while the EMT was driving the ambulance to the hospital, so there was a need for someone to perform compressions. Because there was so much chaos, the fire photographer, who happened to be a retired firefighter from another local town, stepped up to help. Our department policy forbids HFD personnel from operating the ambulance due to liability issues.[26]

Another detail was when the IC was finishing his radio personnel accountability report (PAR) callout immediately following the removal of FF Bell, one of the tactical unit's personnel failed to respond to his accountability request via the radio. Several repeat PAR calls went over the incident channel radio airwaves with no response. The IC immediately issued a new Mayday for the

---

26. Hartford Fire Department, *Administrative Manual—Department Directives.*

now missing firefighter. As the command post attempted to determine where this individual was last seen from his officer, the relieving sound of "He's by the rehab bus" came over the radio. The firefighter had taken off his turnout coat and left it by his apparatus, then walked down the street to the rehab unit to get some water and take a break. Left behind within his turnout coat pocket was his portable radio. When the PAR callout was initiated, he did not have his portable radio with him, and subsequently did not hear the request for PAR. Nobody standing next to him acknowledged the callout at the time either, most likely due to that fact they were all in shock from what just had transpired moments earlier. That is no excuse, but given the immediate circumstances, it's understandable. All remaining firefighters on scene were accounted for, and the IC regrouped several companies and completed suppression evolutions.

Yet another unexpected byproduct of this horrific incident was that the local media, starving for information to report to the public, created their own narrative in the wake of missing, incomplete, or mediocre details. The TV and print media wove a story of conjecture as to what transpired based on a false narrative created from bad intel, hearsay, rumor, manipulated audio tapes, and others badmouthing and blame-laying to protect themselves and their interests both within and outside of the department. Instead of the chief of the department creating a positive narrative complete with factual and relevant information known at the time, the media machine fed off the wound of the department. The infighting and backbiting escalated within the department and this whole incident became one big kick in the groin for all of us above and beyond losing one of our brothers and the inherent errors made to that end.

One major issue for me became abundantly clear in the immediate aftermath of the incident. As the IC and I were standing alone at the command post during the overhaul phase of this fire after the chaos had settled down, the police chief and his assistants quietly walked up to us. It is never a good sign when those guys want to talk to you. The police chief then stated in a very sympathetic and tender tone, his voice low so no one else could hear, "I'm very sorry, it looks like it's going to be a 10-45!" That's the Hartford Police radio code for a fatality.[27] I remember instantly feeling my stomach wrench and having that full body sinking feeling of blood rushing out of my head. I was in a serious state of physical and mental psychogenic shock right then and fought the crushing urge to sit down on the ground, my foundation weakened by overwhelming grief, but I could not display any emotions or react in any way to illuminate any negativity to the troops on scene. "They cannot learn

---

27. "Hartford Police Department [Radio Communication] Codes," 2014, https://www.bearcat1.com/radioct.htm.

this information here and now," I immediately thought to myself. The IC agreed. Was I supposed to gather everyone together and break the news on scene, then lose whatever composure and control that remained after what they had just endured and witnessed? Was that a sound, responsible, cogent decision to put this horrific information on these crews here and now? No way could I do that to my people, my fellow brothers and sisters. There was a better, more dignified, and more respectable way of informing them that their brother firefighter had fallen. The order was given that everyone who responded on the first-alarm assignment would report to our department training division down the street upon leaving the scene. The second-alarm assignment would remain on scene for any further work that was deemed necessary. The senior officer remaining on scene would take command.

The Notification

Once we gathered all the first-alarm assignment personnel into the training division main room on the second floor, I, along with the IC, stood in the hallway outside the room waiting for the chief of department to arrive. Instead, the acting assistant chief arrived, told us that the chief was busy with the media and was not going to show, and left to return to headquarters to be with the chief. That left the two of us on our own. Never in a million years did I think I would be in this position, not only as a witness to the LODD, but also compelled to make the resolute and agonizing notification to the firefighters whom I not only led but also admired and respected as firefighters and fellow brethren. The IC and I returned to the main classroom where there were approximately 50 firefighters and officers seated at the rows of desks throughout the classroom. They were obviously distressed, disheveled, and had a pretty good understanding that something bad had transpired, and being in that room at that time surely confirmed it. We both walked into the room and closed the door behind us, to provide privacy and to isolate the group from any outside interference and interrupters. I stood in the front of the room and the IC stood off to my left, just behind me. I took a minute or two to scan the room and look into each firefighter's eyes, row by row, desk by desk. Dead silence permeated the air. This was clearly an awkward moment for many because my demeanor was not of my normal tone, and the anticipation was as thick as molasses. After looking into the eyes and the somber faces of everyone in that room, I calmly said with fortitude and reverence, "We lost a man tonight!" The room erupted with emotion and reaction as if I had tossed a stick of dynamite into the classroom. It was with all my strength, conviction, and grit that I held my emotions in check and stood tall and firm as their leader, their mentor, and their friend. I have never been so torn up mentally and physically as I was at this exact moment. Some screamed and yelled, some wept both openly and quietly, some cursed and swore, and others sat still, in quiet disbelief and shock. At that moment in time, these firefighters were the embodiment of true anguish, the level and scale as defined by author Brené Brown as an unbearable and traumatic swirl of shock, incredulity, grief, and powerlessness.[28]

All I could think about was the fact that if they had to be notified of a brother firefighter's death, at the very least it came from me and not by some other means or by someone else. Deep down in my heart and mind I believed that at this moment in time during this horrific situation, I somehow was meant to do this, having always adopted the attitude of the leader of the shift, of the members, and of the department. I don't think if the chief of department or anyone else made that notification that those firefighters would have had the willingness to unleash their emotions and react with true passion and dispair and ultimately find humanity and grace in all of this. This situation could have happened to any other chief officer in the department on any other shift, but why was it happening to me? The room began to simmer after about a minute and I then informed them of the resources available to them if they wanted to reach out, either to the Employee Assistance Program or to the department's Chaplain Corps.[29] Then I informed them that when they returned to their respective firehouses, they would be allowed to go home and that an overtime firefighter would take their place for the remainder of the shift. I turned and exited the room, went downstairs to my sports utility vehicle (SUV) with my aide and I returned to my firehouse downtown. The 10-minute ride back was silent. There were no more words to be spoken.

As we were rounding the corner on the approach to the firehouse, I remarked on how many cars were both parked in front of the firehouse and on the street. The streets were clogged with parked cars, and we could barely get into the firehouse. The aide tapped the overhead door button and backed the SUV slowly into the firehouse, carefully maneuvering between the sea of abandoned cars and pickup trucks. There must have been at least 75 or more off-duty firefighters packed into that firehouse, milling around, trying to hear the latest info, attempting to make arrangements and plans, and just to be there to show support to each other and help in any way they could. I was overwhelmed with emotion and began to tear up, my mind already in shut-down mode, my body strained and weak. Their support for me was overwhelming, as they just calmly

---

28. Brené Brown, *Atlas of the Heart: Mapping Meaningful Connection and the Language of Human Experience* (Random House, 2022).
29. Hartford Fire Department, *Administrative Manual—Department Directives.*

helped me empty out my gear from the Red Car (chief's SUV). I was obviously off duty now. A few guys approached me and asked what had happened, but I was still in a state of shock having both dealt with the fire and the notification I just made. I just kept mumbling, "I don't know, I don't know what happened. I know what I saw and what I did but why?" Some approached me with plastic cups half full of whiskey or scotch, in a brotherly and sympathetic attempt to calm me and to show their unwavering support. I was in no position to partake, and I was still in uniform inside the firehouse. I went upstairs to my office, and when I walked in, it looked like a war room, with papers everywhere and two of my chief officer colleagues on the phones and at the desk. They were trying to piece back together the remainder of the shift with overtime and callback personnel and to manage the remaining companies in the city and those left at the fire scene. They acknowledged my presence in the room but were simply too busy to stop and ask what had transpired, having probably already been briefed, and maybe out of respect for me at that moment. These are the guys I work with—my chief officer peers, who at that moment were not only themselves grieving for what transpired but were heartbroken for me as well. I gathered my personal stuff, stripped my bunk, and thanked them for doing what they did best: being there for each other and the department. I went back downstairs and amongst the crowded firehouse full of firefighters, my oldest son was there waiting for me to come downstairs. He had been on the job in Hartford for around 8 years by now and was called at home and informed of the situation. He took my keys and proceeded to drive me home in my truck while another firefighter followed to facilitate a ride back to retrieve his personal vehicle. I was in no shape to drive, my head reeling from the previous events and my emotions in tatters. When we pulled up to my house around 2:00 a.m., my wife was waiting outside on the stairs leading into the house. As I walked up, she embraced me gently and we went inside.

The reason why I cite this anecdote is not readily clear until illuminated: true leadership is not just exclusive to the broad strokes of policy, strategy, and tactics, but likewise found in the subtle, nuanced minutia, as well as in the courage found in all the things that are hard to do.

The Impact

The entire department was impacted by this event. The city had lost a firefighter, and the community had lost a hero. The outpouring of support from all ends of the city was overwhelming at times in that the community, too, was grieving for our loss. Firehouses throughout the entire city were inundated with food, baked goods, cards of sympathy, and visitors wishing to show their gratitude and kindness with words of support and condolence. Time after time, when companies responded for calls of service, the people at these incidents showered the crews with praise and support, and this went on for weeks after the fatal incident.

The brotherhood and sisterhood of our entire department was affected by this tragic event (fig. 8–16). No one was left untouched by what transpired. There was peer support throughout the department, both for those who were at the scene as well as for those who had nothing to do with the incident. Calls came in from retired firefighters and from firefighters all over the country wanting to assist and show their support. I returned to work the very next scheduled shift feeling compelled to be with the members on my shift not only to show my support for them, but also because we were all grieving and emotionally raw, and I believed being with them was the best place at that time. It was business as usual the next shift and I demanded of myself to project strength and guidance for the firefighters and officers, for they truly were the most important people in this time of sorrow and need, and I asked the on-duty officers to do the same in the weeks to come.

FIG. 8–16. The brotherhood, with me on the far left, in 1993 (Photo courtesy of Joe Marino)

What's Left

This tragedy will no doubt have a ripple effect that will be felt for years to come. The outpouring of support, the investigations and reports written, the needed changes within, and the healed wounds are all the result of one fateful night at one disastrous incident. For me, this fire sowed seeds of doubt. While responding to another report of a building fire several days later, I began to experience anxiety and became extremely nervous while responding and operating at this next working fire incident. Trepidation had creeped into my head and the heads of other firefighters because of the dreaded anticipation of another bad event happening. Although having difficulty with composure and concentration, I was able to work through these feelings by being with my crews and talking to them and especially my aide, who although was not with me the night of the fatal fire, was none the less instrumental in helping me and others get back on track with our emotions and composure. Guilt is a normal reaction to something of this magnitude, especially when you are in a leadership capacity. It is something you must process and work through to become stronger, better prepared, and not only survive, but thrive as well (fig. 8–17).

There is also plenty of fault to go around. With the investigations conducted and the reports written, there is blame and responsibility everywhere, at all levels. As a leader, you must be prepared to answer to these pointed assessments. There is no other way to succeed in this endeavor than to learn from your mistakes and grow. The way back for me, among several ways of processing and handling all the overwhelming byproducts of this fateful event, was to

FIG. 8–17. The funeral for FF Bell (Photo courtesy of Pat Dooley)

contact several of my peers who'd had similar experiences at incidents. I reached out to chief officers who had recently suffered a LODD while on their respective shifts in departments both locally and out of state. Hearing the words of experience from someone in a similar situation was comforting, uplifting, and gave me the internal strength needed to move forward and be a better firefighter, chief officer, and person. There were multiple official reports written about this incident that are available for public consumption. Unfortunately, in this type of situation, these are not chock full of accolades and intense moments of heroism. Instead, they are a stark memorialization of the failures and shortcomings that ultimately led to the loss of a human life, one of our brother firefighters.

On May 2, 1988, my first day of recruit training as a Hartford firefighter, we were corralled into a small classroom in our training division and sat to be photographed for our department identification cards. The photographer took two photos of us, one with a department ballcap which we were just issued, and one without. As you can see in these two separate and distinct identification cards (fig. 8–18), I did not even have on a uniform shirt yet. The date on the card is when it was issued.

I asked the nice lady taking the photos why two different pictures, with and without a hat. She casually replied with a somewhat sarcastic and indifferent tone, "The one with the hat is what they give the newspaper when you get killed!" I was both stunned and floored! Get killed?! Is that what they have in store for us? How often does that happen! What did I get myself into here?

Fast forward to the mid-2000s in my department. I was already an established captain at Engine Company 10 in the city's Little Italy neighborhood in the south end when the department executive officer showed up one day at my double-company firehouse and asked that each of us briefly allow for a photograph to be taken. He told me it was for the chief's office reference, in that there were so many new hires, the chief could not keep track and thought placing a face with a name would help. I told him I knew exactly what the real reason was. We all gathered briefly and had an informal photo taken of our faces. Several years later, after I had been promoted to deputy chief, I volunteered to be detailed to the chief's office as the interim assistant fire chief. It was a great opportunity to learn the administration side of the department as well as to have a direct impact on policy, procedure, and the day-to-day operation of the department. On a shelf in the chief's office was the three-ring binder of all the 8"×10" glossy headshot photos that were taken that day, and on the cover of the binder was the title in bold black letters, "Hartford Fire Department Face Book" (fig. 8–19). Sure, it helped the chief put a name to a face, but the real reason it existed was that it held the photos that would be distributed to the media in the event someone within the department lost their life (fig. 8–20).

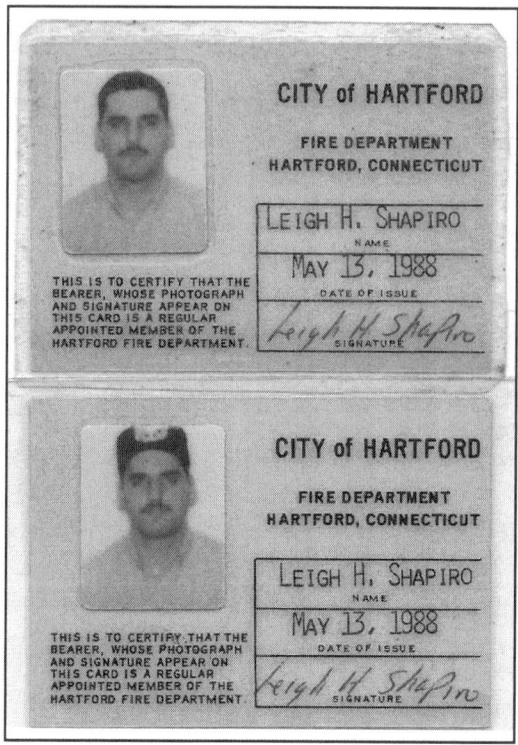

FIG. 8–18. These are the two ID cards issued to me at the beginning of my career in May of 1988. The photos were taken my first day on May 2, and the cards were issued on May 13. As you can see, I didn't even have a uniform yet, just that hat.

Chief of Department Edward F. Croker of the FDNY said it most profoundly when he gave this statement upon the death of a deputy chief and four firefighters in February of 1908:[30]

> *Firemen are going to get killed. When they join the fire department, they face that fact. When a man becomes a fireman, his greatest act of bravery has been accomplished. What he does after that is all in the line of work. They were not thinking of getting killed when they went where death lurked. They went there to put the fire out and got killed. Firefighters do not regard themselves as heroes because they do what the business requires (fig. 8–21).*

---

30. Chief Ed Croker, as quoted in *War* by Sebastian Junger (Twelve, 2011).

8 • The Incident: The Impact of a Fire Service Line-of-Duty Death 203

FIG. 8–19. The official HFD Face Book, as it was affectionately named, containing photos of the entire department's personnel. The three-ring binder is filed alphabetically, and my son and I are shown here in the book. This photo was taken prior to my youngest son Quentin also being hired as a firefighter. It's details like these that hit home and are a bare reminder of how dangerous our mission really is.

FIG. 8–20. FF Kevin L. Bell (Photo courtesy of HFD)

FIG. 8–21. The firefighters who worked directly with FF Bell remove his casket from the apparatus and carry him into the church. Note the firefighter with the bandaged head: he was injured at this fire incident as well. (Photo courtesy of Pat Dooley)

Case Study

Immediately following the removal of the downed firefighter, command initiated a personnel accountability report (PAR) via the radio to determine if all the other firefighters were accounted for. As stated earlier, one firefighter did not respond to the radio call out for PAR and the IC immediately issued a new Mayday for the now missing firefighter. He was ultimately found at the rehab bus on scene and left his portable radio in the pocket of his turnout coat, which was hanging on a grab rail of his apparatus at the far end of the fireground.

Questions

1. Given the inherent chaotic nature of the event at this incident, what policies or training do you believe would have mitigated the second Mayday event?
2. What within the department's culture at the time can you identify and define that you believe led to the second Mayday?
3. Which elements of the NFFF's 16 Firefighter Life Safety Initiatives address leading change in the fire service in direct relation to this LODD incident?[31]

---

31. NFFF, "16 Firefighter Life Safety Initiatives," Everyonegoeshome.com, 2023, https://www.everyonegoeshome.com/16-initiatives/.

9

Final Thoughts

Firefighters know firefighters! We often gauge ourselves and other firefighters as good people, or good firefighters, among other positive accolades and terms of endearment (fig. 9–1). Your reputation as a firefighter is a commodity which you can use to advance your agenda, make positive connections, and move the ball forward in both your professional and personal life. You should be dependable in your actions and involvement, competent in your education and training, trustworthy in your interactions with other firefighters and the public, capable of getting the job done whatever it may take, and reliable in your input and voice. Otherwise, you are simply here for the T-shirt, the bling, and all the other wrong reasons one would be in this profession for reasons other than noble. Take ownership of your experiences and situations. This is

FIG. 9–1. "That's my brother Goddammit!" blurts an injured Lieutenant Stephen McCaffrey while observing his brother and fellow firefighter Brian McCaffrey wield a handline and save the day in the finale scene from the movie *Backdraft*. Even in his physical state, he still can project pride and respect for the brave and heroic firefighters he serves with.* (Screenshot of video clip courtesy of Universal Studios)

*Backdraft, directed by Ron Howard (Universal Studios, 1991).

how we learn, grow, and eventually become the mentor we have often sought out. Be engaged in those situations where you find yourself. Learn from each of them, and don't allow your experiences to be fuel for just another good story or anecdote. What did you learn and from whom, and what changed you if you could identify something worthy? Your contribution to the mission of the fire service is most valuable and can have a profound impact on both yourself, others with you, and our mission overall.

This job has often been compared to combat military service in that we work together, live together, eat and sleep with our brethren, laugh and cry with each other in times of good and bad, and have the common core thread of pride, respect, and the duty to act in a professional manner regardless of personal emotion or opinion. No other job on the planet is as rewarding as the fire service. Think about that for a moment. For the most part, people are happy to see us, especially when they are probably having the worst day of their life. People respect and admire our profession. We can stand in front of a building after a fire and say, "Look what we did—we saved this property" or "We saved a life, or multiple lives." We make a physical, tangible, measurable difference. We protect the public and the property in our community, neighborhood, street, and home. Unfortunately, the police don't always get that same welcome. Usually, the police show up after something has occurred, whereas the fire service shows up when it's happening, and that's what sets us apart in our missions. I have often told those firefighters who I have had the honor of instructing that there is an addendum to the old saying, "The greatest gift a father can give his children is to love their mother," which is, "The greatest gift a fire officer can give to their crew is to spend time with them." Nurture them, mentor them, empower them, and allow them to learn from your training, education, and experiences. Act with compassion and validate those who serve with you. Recognize and embrace those meaningful connections you have made (fig. 9–2).

FIG. 9–2. The look of respect and admiration I often gave to my crews and their efforts (Photo courtesy of Pat Dooley)

It's not about you and your special needs. It's about the mission and what you bring to it! You are the future of your department, and what you do now to prepare for that inevitability will have a profound impact on you, your department, the personnel around you and that come after you, and the fire service. Stay the course and focus your energy on being an agent of change, rather than being a mere bystander who is easily distracted. Our mission and your contribution are too important to be compromised.

Case Study

When we join the fire service in any capacity, whether career or volunteer, there comes a moment when you must look back and reflect on your time, tenure, experiences, and the people you have encountered along the way. I have often felt disappointed and frustrated with those firefighters whom I have had the honor of serving with who were great at what they did but simply took a left turn when they retired and did not give back to the fire service that which they experienced and learned throughout their tenure. In my opinion, it's very judgmental to feel this way, but I can't help it. You were so good at what you did, so why would you not want to at least share your knowledge, training, education, and experience with others who are not only thirsty for but also in real need of the very thing you possess? This is the true learning within the fire service. When I retired, I really didn't think anyone would be remotely interested in what I had to offer. I was wrong. Today, my experience and knowledge are actively sought because it's real, true, and relevant. As the Terminator stated, "I'm old, not obsolete!"[1] Retired firefighters and fire officers should be looked upon not as washed up, but as sources of wisdom (fig. 9–3). I have

FIG. 9–3. The career challenge coin I received from the chief of department on my final shift

---

1. *Terminator Genisys*, directed by Alan Taylor (Paramount Studios, 2015).

learned through much internal soul-searching that to be most effective as a retired firefighter, you must quantify your impact and be visible to those in search of mentorship, because your best ability is availability![2]

Questions

1. What steps can be initiated to collect and document your knowledge, accomplishments, and milestones throughout your career in order to pass along that body of work to the next generation of firefighters and fire officers?
2. What consideration, if any, have you given to your contribution and legacy you will leave for future generations to propel the fire service mission forward?

---

2. Although this quote can be attributed to both Bill Parcells and George Kittle, in its context and how it was stated, I am referring to a quote from National Football League player Todd Gurley in the HBO TV series *The Shop* season 1 episode 3.

Index

A

AC (assistant fire chief) 32
accountability
    of operations 124
    system 174
acting
    assistant chief 195
    district chief 119
    driver 164
    officer 37
action, corrective 44–45
active fire 114
    protection system 134
adapting tactics 157
address, incident 123
adjutant 125
administration 101
    fire 46, 51
administrative chiefs 124
aide to the chief 121
alert systems for firehouses 120
all-hazards response department 100
all-in mentality 152
alternative plan 157
anxiety, exam 65
apparatus 62, 100
    ladder 89
assessment center 60
assistant fire chief (AC) 32
attack
    defensive 173
    line 79, 125
    offensive 169
    transitional 160
attitude, positive 96
authority, temporary 38
autopsy protocol for firefighters 191
awareness
    keen 114
    peripheral 113
    situational 11, 113

B

bailout 97
barrier to communication 45
basic
    officer procedure 46
    template 77
battalion chief (BC) 177
    Mike O'Hallorhan 177
BDA. *See* before, during, and after (BDA)
Beacon Street Fire 180
bedroom fire 4
before, during, and after (BDA) 43
    model 52
behavior, human 142, 154
    fire-related 149
Bell, Kevin 180
big picture 30, 97
bleach spill 66
book knowledge 101
Boston firefighters 180
Boston's back bay 180
bottom line 43

Bradberry, Travis 28
braggart 93
breath, first 49
Brown, Brené 196
building, high-rise 133

C

Captain Chaos 111
captain (company) 42
carcinogens 159
career fire departments 162
career sabotage 28
caretaker 34
Carpenter Street incident 172
catch-up, incident 156
CBA. *See* collective bargaining agreement (CBA)
certifications, mandated 99
chain of command 51
change and firefighters 29
chaplain corps 164, 196
charged handline 4
Chief of Department Edward F. Croker 202
chiefs
    administrative 124
    aide 121
    deputy 61
    Kastros, Anthony 61
    officer 10, 119, 197
    training 32
city taxpayers 63
classes, training 99
Clay, Zandra 48
clinical impression 129
coaching 7
cognitive comprehension 142, 153
coherent communication 96
cohesive intervention 109
cohesiveness, group 146
collective bargaining agreement (CBA) 12, 69, 101–103. *See also* labor contract
college degrees 99

Combs, Paul 118
comfort level 113–114
command
    board 121, 124, 127
    chain of 51
    post 4, 114, 118, 121, 125, 130
    presence 96, 116, 119, 125
    structure 169
commander (tour) 23, 76
committee, health and safety 117
communication
    barrier 45
    coherent 96
    constant 126
    of rationale 116
    poor 81
    reliable 172
company
    captain 42
    heavy rescue 122
    officer 157
company, ladder 5, 116
    first-due 122
compartmentalizing tasks 128
complaint
    sexual harassment 48
    soft 50
compounding information 119
comprehension, cognitive 142, 153
computer statistics (COMPSTAT) 32
Connecticut
    Fire Marshal certification 11
    Fire Safety Code 91
    Forensic Science Laboratory 183
    Statewide Honor Guard 192
considerations, primary 114
constant communication 126
construction, legacy 122, 171
contemporary firefighting 159
contract, labor 60. *See also* collective bargaining agreement (CBA)
control-freak mode 38
corrective action 44–45
courses, refresher 99

court
  labor 38
  scrutiny 52
crew
  engine 97
  fire 110, 146
  status 88
critical thinking 16, 96, 99
Croker, Edward F. 202
cultural differences 45
curriculum vitae (CV) 59–60

D

dance card method 121–124, 127
Daubert standard method 160
day-to-day dealings 35
decision-making process 16
decompress 65
de facto safety officer 72
defensive attack 173
deficiencies, educational 45
definitive policies 101
degrees, college 99
department
  all-hazards response 100
  directives 90, 101, 169
  licenses and inspections 132
  policy 62, 87
deprivation, sleep 144
deputy chief 61
  district 125
  promotion 202
differences, cultural 45
diligence, due 114
directives
  department 90, 101, 169
  operating 30
  violating 101
director, emergency management 99
discipline
  progressive 46
  punitive 46
district deputy chief 125
divisions, training 99

documenting information 43–44
do-not-enter method 158
double-company firehouse 201
downed firefighter 181
drill school 53
due diligence 114
Dumont, Eugene 115

E

EAP. *See* employee assistance program (EAP)
educational deficiencies 45
ego 29
eight-step process 77
elements, structural 171
eliminating risk 162
emergency management director 99
emergency medical services (EMS) 1, 5
  call 31–32
  template 78
emergency medical technician (EMT) 51, 129
emergency operations center 32
emergency room (ER) 5
emotional intelligence 30, 33, 97
employee assistance program (EAP) 5, 164, 196
empowerment 23–24
EMS. *See* emergency medical services (EMS)
EMT (emergency medical technician) 51, 129
engine company
  captain 62
  officer 125
  10 163
engine crew 97
enlightenment 97
equipment 100
ER (emergency room) 5
evolutions, suppression 149
exams
  anxiety 65
  fire marshal 62

exams (*continued*)
    lieutenants 13
    promotional 60
exercise, role reversal 45
experience, prior 78
exterior
    attack line 79
    extinguishment method 159
    operations-only method 160

F

FAA (Federal Aviation Administration) 142–143
facebook 10
factor, human 98, 159
failure 43, 66
family impact 47
fatal incidents 114
FDIC (Fire Department Instructors Conference) 61
Federal Aviation Administration (FAA) 142–143
Federal Q Siren 2
feelings, physical 119
FF. *See* firefighters (FF)
Fire and Emergency Services Higher Education 100
Fire Department Instructors Conference (FDIC) 61
fire departments
    career 162
    Sacramento (CA) 61
fire emergency reactions 142
Fire Engineering 61
firefighters (FF) 178
    age 144
    autopsy protocol 191
    behavior 154
    change and 29
    Clay, Zandra 48
    downed 181
    female 48
    Hartford 200
    Jones 50
    new 97
    off-duty 197
    Ortiz, Maria 48
    performance 144
    probationary (probie) 42, 53
    reputation 207
    retired 209
    survival 154
firefighting
    contemporary 159
    future 158
    old-school 156
fireground 90, 157
firehouse 1, 198
    alert system 120
    double-company 201
fire marshal 99
    oral exam 62
    school 59
fire officer 9–13, 92–93, 97, 99, 209
    career challenge 46
    job description 70–71
    reputation 47
fire-related human behavior 149
fires 62
    academy training 94
    active 114
    administration 46, 51
    Beacon Street 180
    bedroom 4
    crew 110, 146
    food-on-the-stove 62
    incidents 63
    industrial complex 116
    large brush 105
    Notre Dame cathedral 159
    physical setting 147
    science 100, 160, 162
    Station Nightclub 149
    working 3, 158
Fire Safety and Building Codes 91
fire service 57, 99, 139, 156, 159, 181
    agency 100
    gender and 147

incidents 94
mission 208
modern 48
oath 93
student 11
trust and 137
first breath 49. *See also* sexual harassment complaint
first-due ladder company 122
Fischer, Thomas A. 178
formalized structure 146
freelancing 90
frequent training 88, 101
future of firefighting 158

G

gabled roof 171
gender and fire service 147
golden rule 96
Goldstein, Arnold 115
grooming 55
group cohesiveness 146

H

handline, charged 4
hands-on world 109
Harley rider 29
Hartford
   Courant 179
   firefighter 200
   fire marshal 6
   plaster hook 76
   police 194
Hartford Fire Department (HFD) ix, 90, 120, 135, 186
hazmat incident 77, 131
health and safety committee 117
heavy rescue company 122
heuristic 152
HFD. *See* Hartford Fire Department (HFD)
high-rise building 133
   operations 133
high school, Marjory Stoneman Douglas 162

hiring process 57
Holden, William 177
hostile radio traffic 119
human
   factor 98, 159
   interactive situations 94
   mentality 33
   nature 145
   nervous system 181
human behavior 142, 154
   fire-related 149
humility 37
hydrant personnel 171

I

IAFC (International Association of Fire Chiefs) 191
IAP. *See* incident action plan (IAP)
IC. *See* incident commander (IC)
IFSTA (International Fire Service Training Association) 117
impact, family 47
impression, clinical 129
incident action plan (IAP) 129–130, 159, 167
incident commander (IC) 3, 72, 85, 109, 172
   acting 88, 113, 124, 135, 167
   responsibilities 82, 116, 119, 138–139, 157, 181
incidents
   address 123
   Carpenter Street 172
   catch-up 156
   command management at 119
   fatal 114
   fire service 94
   hazmat 77, 131
   management 113, 128
   multiple-alarm 124
   successful fire 62
individual morality 103
industrial complex fire 116

information
  compounding 119
  documenting 43–44
initial
  action plan 155
  alarm assignment 124
  attack line 171
inspections and licenses department 132
instilling
  confidence 38
  pride 39
  temporary authority 38
institutional knowledge 101
Insurance Services Office Class II 63
intellectual clarity 154
intelligence, emotional 30, 33, 97
interacting with the public 41
interactive situations, human 94
intercom system 120
interior
  attack 169
  progress 88
  suppression evolution 158
International Association of Fire Chiefs (IAFC) 191
International Association of Fire Fighters 102
International Fire Service Training Association (IFSTA) 117
intervention, cohesive 109
intrinsic behaviors 96
issues, personnel 27
issuing orders 116

J

jargon 45
job performance requirements (JPRs) 105–106

K

Kastros, Anthony 61
keen awareness 114
knowledge, institutional 101
knowledge, skills, and abilities (KSAs) 17, 22

L

labor
  contract 60. *See also* collective bargaining agreement (CBA)
  court 38
ladder apparatus 89
ladder company 5, 116
  first-due 122
  second-due 122
large brush fire 105
law firm 50
leadership 12, 97
  personal 22
  transactional 22
  transformational 22–25
legacy construction 122, 171
legislator, state 63
liability, minimizing 99
licenses and inspections department 132
Liebowitz, Fran 179
Lieutenant Arnold Goldstein 115
lieutenants exam 13
line
  attack 79, 125
  bottom 43
  exterior attack 79
line-of-duty death (LODD) 10, 79, 147, 178
Little Italy 201
local media 194
LODD. *See* line-of-duty death (LODD)
lucidity, psychological 154

M

management 95
  by walking around 127, 138
  incident 113, 128
mandated certifications 99
mantra 135
margin of safety 149
Marjory Stoneman Douglas high school 162
mass shooting 162
Mayday 81, 154

McCammon, Ian 152
McQueen, Steve 177
mechanics 95
media, local 194
memory, muscle 62
mentality, human 33
mental strength 83
mentoring 7, 18–19, 22, 56, 210
    junior officers 128
message (radio) 115
methods
    dance card 121–124, 127
    Daubert standard 160
    do-not-enter 158
    exterior extinguishment 159
    exterior operations-only 160
    management by walking around 127, 138
    transitional knock-back 160
micromanaging 116
military 159, 208
minimizing liability 99
modern fire service 48
Moltke, Helmuth von 155
morality, individual 103
multiple-alarm incidents 124
muscle memory 62

N

National Aeronautics and Space Administration's (NASA) 124
National Fallen Firefighters Foundation (NFFF) 190
National Fire Academy 100
National Fire Incident Reporting System 121
National Fire Protection Association (NFPA) 15–16, 62
    Guide for Fire and Explosion Investigations 141
National Institute for Occupational Safety and Health (NIOSH) 153
nature, human 145
navigating negativity 28

nervous system, human 181
new firefighters 97
New York Fire Department 202
NFFF (National Fallen Firefighters Foundation) 190
NFPA. *See* National Fire Protection Association (NFPA)
NIOSH (National Institute for Occupational Safety and Health) 153
Notre Dame cathedral fire 159

O

oath, fire service 93
off-duty firefighter 197
offensive attack 169
officer procedure, basic 46
officers 11, 26
    chief 10, 119, 197
    company 157
    resentment 41
    training 99
O'Hallorhan, Mike 177
old-school firefighting 156
omitting steps 64
ongoing process 114
operations
    accountability 124
    command tone and expectations 27
    directive 30
    high-rise 133
    water supply 171
operators, pump 171
oral
    exams 61–62
    reprimand 46
organization
    paramilitary 36
    scalar 36
Ortiz, Maria 48
overhaul
    and salvage 113
    phase 114
ownership, taking 207

P

PAR. *See* personnel accountability report (PAR)
parade, St. Patrick's Day 75
paradigm shift 24, 159, 174
paralysis, task 113
paramilitary organization 36
Parcells, Bill 25
Paris, France 159
Parkland, Florida 162
passive fire protection system 134
pension qualifications 47
perception, sensory 151
peripheral awareness 113
perishables, replenishment 100
personal
 alert safety system device 88
 leadership 22
 style 36
personal protective equipment (PPE) 156, 167
personnel accountability report (PAR) 4, 169, 194
personnel, hydrant 171
personnel issues 27
phase, overhaul 114
physical
 feelings 119
 setting 147
pillars, principle 99
plan, alternative 157
planning, succession 55
policies, definitive 101
policy, department 62, 87
poor communication 81
portable radio 185
positive attitude 96
post-incident analysis 27, 167
PPE (personal protective equipment) 156, 167
pre-incident planning 190
preparing subordinates 55
presidents
 Reagan, Ronald 138
 Roosevelt, Franklin Delano 46
 union 117
primary
 considerations 114
 search 122
principle pillars 99
prior experience 78
prison yard story 34
private sector 57
probationary firefighter (probie) 42, 53
problem
 employee 35
 solver 63
 solving skills 97
processes
 decision-making 16
 eight-step 77
 hiring 57
 ongoing 114
professional 12
 development model 100
progressive discipline 46
promotion 21, 56
 announcement 57
 deputy chief 202
 exam 60
 opportunities 56, 65
psychological lucidity 154
public interaction 41
pump operators 171
punitive discipline 46

Q

Q Siren, Federal 2
qualifications, pension 47

R

radio
 message 115
 portable 185
 reports 119, 127
 system communication 81
 transmissions 88
rapid egress 87
rapid intervention team (RIT) 4, 87–89
rationale, communication of 116
Reagan, Ronald 138

rebound success 66
RECEO VS (rescue, exposures, confinement, extinguish, overhaul, ventilation, and salvage) 77, 130
recertification 99
red car 19–21, 111, 121, 197
reevaluation, situation 155
reflex training 153
refresher courses 99
reliable communication 172
replenishing perishables 100
reputational value 138
requirements, training 99
rescue, exposures, confinement, extinguish, overhaul, ventilation, and salvage (RECEO VS) 77, 130
resentment, officer 41
response, startle 142
resume 57
retired firefighters 209
reward versus risk 85, 103
Rhode Island 149
Rhodes scholar 162
rider, Harley 29
rip-and-run sheet 121
ripple effect 199
risk
   eliminating 162
   versus reward 85, 103
RIT (rapid intervention team) 4, 87–89
role reversal exercise 45
roof, gabled 171
rooms, watch 121
Roosevelt, Franklin Delano 46
Roosevelt, Theodore Jr. 157
rope tag line 130
rule, golden 96

S

sabotage, career 28
Sacramento California Fire Department 61
safety and health committee 117
safety, margin of 149
safety officer 3, 73
   de facto 72
Saint-Exupery, Antoine de 55
St. Patrick's Day parade 75
salvage and overhaul 113
San Luis Obispo, California 160
scalar organization 36
SCBA. *See* self-contained breathing apparatus (SCBA)
school
   drill 53
   fire marshal 59
science, fire 100, 160, 162
scientific method 55
scrutiny, court 52
search, primary 122
second-due ladder company 122
self-contained breathing apparatus (SCBA) 5, 145, 153
sensory perception 151
setting, physical 147
sexual harassment complaint 48. *See also* first breath
sheet, rip-and-run 121
shift, paradigm 24, 159, 174
shooting, mass 162
situational awareness 11, 113
situation reevaluation 155
size-up, 360-degree 124
size-up, set up, support method 128
sleep deprivation 144
soft complaint 50
SOPs (standard operating procedures) 26
Soviet Union 124, 138
spill, bleach 66
sports utility vehicle (SUV) 106, 196
sprinkler and standpipe system 149
standard operating procedures (SOPs) 26
standpipe and sprinkler system 149
startle response 142
state legislator 63
State of Connecticut Office of the Medical Examiner 191
Station Nightclub fire 149

status, crew 88
steps, omitting 64
stories
    prison yard 34
    war 26
Street Boss 113
strength, mental 83
structure
    elements 171
    formalized 146
    unprotected 149
student, fire service 11
style, personal 36
success, fire incident 63
succession
    development 97
    planning 55
success, rebound 66
support system 100
suppression evolutions 149
suspension days 47
SUV (sports utility vehicle) 106, 196
synergy 17, 80
systems
    accountability 174
    active fire protection 134
    firehouse alert 120
    intercom 120
    passive fire protection 134
    sprinkler and standpipe 149
    standpipe and sprinkler 149
    support 100
    vocalarm 120

T

Tac-1 1
tactical unit 122
tactics, adapting 157
tag line, rope 130
taking ownership 207
tasks
    compartmentalizing 128
    paralysis 113
taxpayers, city 63

teamwork 78
template, basic 77
temporary authority 38
thinking, critical 16, 96, 99
360-degree size-up 124
timewaster (TW) 113, 117
tour commander 23, 76
Towering Inferno 148, 177
training 99
    chief 32
    classes 99
    divisions 99
    fire academy 94
    frequent 88, 101
    officer 99
    reflex 153
    requirements 99
transactional leadership 22
transformational leadership 22–25
transitional attack 160
transitional knock-back method 160
TW (timewaster) 113, 117
Tzu, Lao 25, 97

U

Underwriters Laboratories (UL) 160
union president 117
unit, tactical 122
unprotected structure 149
U.S. Department of Education 57

V

value, reputational 138
violating directives 101
vocalarm system 120

W

war stories 26
watch rooms 121
water supply operations 171
wisdom 21, 97
working fire 3, 158
world, hands-on 109
World War II 157